Coloring of Plastics

Theory and Practice

Coloring of Plastics

Theory and Practice

Mukhtar Ahmed

VNR VAN NOSTRAND REINHOLD COMPANY
NEW YORK CINCINNATI ATLANTA DALLAS SAN FRANCISCO
LONDON TORONTO MELBOURNE

Van Nostrand Reinhold Company Regional Offices:
New York Cincinnati Atlanta Dallas San Francisco

Van Nostrand Reinhold Company International Offices:
London Toronto Melbourne

Library of Congress Catalog Card Number: 78-26186
ISBN: 0-442-20267-9

Manufactured in the United States of America

Published by Van Nostrand Reinhold Company
135 West 50th Street, New York. N.Y. 10020

Published simultaneously in Canada by Van Nostrand Reinhold Ltd.

15 14 13 12 11 10 9 8 7 6 5 4 3 2 1

Library of Congress Cataloging in Publication Data

Ahmed, Mukhtar.
Coloring of plastics.

Includes index.
1. Plastics – Coloring. I. Title.
TP1170.A35 668.4'1 78-26186
ISBN 0-442-20267-9

Introduction

Color is an integral part of consumer products, especially those made of plastics. Originally, color was used for masking the natural coloration or discoloration of thermoset resins. With the availability of essentially colorless thermoplastics, this function of color has decreased in importance. The coloration of plastics is now not only an essential tool for sales promotion but also is an important factor in product design. The markets of interest to the plastics industry have constantly indicated that the ability to color plastics is essential to the continued growth and promotion of new products. In fact, color and appearance have often been used as primary vehicles to penetrate various consumer and industrial markets in the recent past.

Considering its importance, it seems strange that coloration has never been regarded as a particularly vital segment of technological activities in the plastics industry. Coloring problems have generally been treated as common headaches of extrusion and injection molding operations; they have not been worthy of the attention dyeing and printing receive in the textile industry. The general lack of concern about the coloration of plastics is indicated by the fact that not a single book has been written in the English language discussing the problems and prospects of this vital activity of plastics industry. It is only recently that the Society of Plastics Engineers (SPE) began to organize regional technical conferences on this topic.

The need for the dissemination of information and the exchange of ideas on coloring of plastics was there from the start. The current heightened interest is a result of the introduction of very intensely colored organic pigments to the plastics industry, the availability of high melting plastics, the need for high through-put machinery, the maximization of efficiency of the plastic fabricating equipment, the demand for brighter and high performance colors, and cost cutting endeavors in plastics fabrication. In addition, due to an increased scale of production and assembly line operations of the industries utilizing the colored plastics products, the control of color and appearance has become critical from item to item and from batch to batch. This has necessitated the application of stricter quality control measures, and indicated the need for instrumentation in plastics coloring operations. As though these pressures were not enough, the increasing concern of various agencies of the government for public safety compelled the plastics color technologist to acquire more basic knowledge about the materials he utilizes and the way they would behave under use conditions. There was evidently a need to consolidate this information.

Basically there are two methods of coloration of plastics: surface coloration (painting, printing, and dyeing) and mass coloration. The proportion of plastics colored by the first method is very small: the major part of plastics articles are colored *en masse.* This book is therefore devoted mostly to mass coloration of plastics (pigmentation), although surface decoration has not been neglected.

The problems of coloring of plastics are many: some defy solution, some have easy answers. While this book does not provide an exhaustive examination of every aspect of coloring technology for plastics, it is hoped that it will serve as a basic guide to this increasingly important area of plastics technology.

M. AHMED

ACKNOWLEDGMENTS

This book is heavily based on the material presented by various experts in recently held ANTECs and RECTECs of the Society of Plastics Engineers. This contribution is thankfully acknowledged without giving specific names and without enlisting a large number of specifics. In addition, the following non-SPE sources are gratefully acknowledged of which a liberal use has been made in the preparation of this book:

Bilmeyer, F. W. and Salzman, Max *Principles of Color Technology.* New York John Wiley: (1966), for the introductory material of Chapter 1, 2 and 7.

Rieve, T. B. and Dills, W. L. "Principles of Pigment Dispersion in Plastics." *J. Color Appearance* **I** pp. 25–29 (1971), for theoretical discussion of Chapter 6.

Schoengood, A. A "Compounding." *SPE J.* (Feb. 1973), for most of the material covered in Chapter 8.

Kut, S. "Painting." *Modern Plastic Encyclopedia.* New York: McGraw Hill, for section on "Painting" in Chapter 9.

Skeels, A. N., Jr. *Modern Plastic Encyclopedia.* New York: McGraw Hill, for section on "Printing" in Chapter 9.

Ackroyd, P. "The Mass Coloration of Man-made Fibers." *Rev. Prog. Color.* **5**, p. 86 (1975), for the introductory material of Chapter 11.

Ahmed, M. *Polypropylene Fibers – Science & Technology.* Elsevier, Amsterdam, the Netherlands, for the material on polypropylene in Chapter 11.

Contents

Contents

Coloring of Plastics

Theory and Practice

1 Color and Appearance

INTRODUCTION

Color is one small part of the much larger subject of appearance, i.e., the way things look. The variables of appearance include, apart from color, such things as glossiness or matness, opacity or transparency, smoothness or roughness, and metallic reflection or diffusing surfaces. If these appearance characteristics are eliminated, what remains is absorption of light by the object in such a way that some colors are removed from the light and some are reflected or transmitted. The color we see in an object, however, is not in the object but in our brains. This brings us to a primary definition: *color* is our brain's perception of light that has been modified by the absorption of part of its rays.

The committee on colorimetry of the Optical Society of America has defined *color* as consisting of "the characteristics of light other than spatial and temporal inhomogeneities, light being that aspect of radiant energy of which a human observer is aware through the visual sensations which arise from the stimulation of the retina." Color, thus defined, is a psychological response to a physical stimulus. In simple terms, color is what we *see* – the result of the physical modification of light by colorants as observed by the human eye and interpreted in the brain. Color perception depends not only on the inherent color characteristics of an object, it is also influenced by temporal, spatial, or appearance aspects, as well as by the surroundings of the object. It is also influenced by the individual factors of past experience and memory, attitude, and individual visual response characteristics.

PHYSICAL BASIS OF COLOR

From the purely physical point of view, the production of color requires three things: a source of light, an object to be illuminated, and an eye and brain to perceive the color. Alternatively, the eye may be replaced by a photosensitive detector and auxiliary equipment which approximates the eye's action in detecting light.

Light can be described by its wavelength, for which the millimicron (mμ) or nanometer (nm) is a convenient unit of length. One nm is 1/1,000,000 millimeter. A balanced, essentially white, light source contains a full range of wavelengths or radiation. The relative insensitivity of the eye limits the visible part of the spectrum to a very narrow band of wavelengths, between about 380 and 750 nm. Each wavelength of visible light stimulates the eye to perceive some particular color, as shown (approximately) in Fig. 1. When light strikes an object, one or more things can happen. The light can pass through essentially unchanged – i.e., it can be *transmitted* through the material, and the object is then called transparent.

At every boundary between two different materials (such as an air and plastic interface, or between two different plastics) light changes its speed. As a result, a small fraction of the light is *reflected.* Surface gloss results when a portion (generally a few percent) of the incident light is reflected specularly: i.e., the incident ray and the reflected ray form equal angles to the reflecting surface in a plane

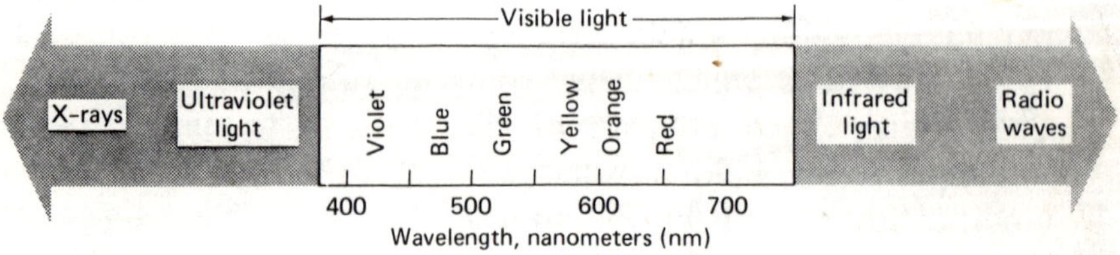

Fig. 1. The visible spectrum and its relation to other kinds of radiation.

perpendicular to the surface. This produces specularly reflected highlights, seen at certain viewing angles or in certain areas of an irregular surface. The observer interprets these highlights in terms of gloss and shape. Highly polished surfaces produce bright, sharp highlights, while microscopic roughness results in a diffuse reflection. In general, regardless of how solid the color, the reflected image is colorless. An exception occurs when a concentration of certain colorants at the air interface causes bronzing, in which reflection itself is colored. In obtaining the desired styling impression, from liveliness to subdued richness of color, control of surface gloss is essential. This can be gained either by controlling the texture of the surface against which the plastic is formed, or by applying a surface coating of the desired balance of gloss-producing resins and dulling extenders.

In addition to being transmitted and reflected, light may also be *absorbed.* If the material absorbs part of the light, it appears colored but is still transparent; if all light is absorbed, the material appears black. The absorption of light is governed by two laws of physics – Lambert's law and Beer's law. The former states that absorption of light at each wavelength is proportional to the thickness of the material. The second law states that the absorption at each wavelength is proportional to the amount of absorbing material (colorant).

Finally, light may be *scattered.* When it passes through matter, if only a part of the light is scattered, the material is said to be translucent. If the scattering is so intense that no light passes through the material, it is said to be opaque. In both cases the color of the material depends on the amount and kind of absorption present. If there is no absorption, a scattering material looks white; otherwise, it looks colored.

Scattering is caused by light falling on small particles whose refractive index is different from that of the surrounding materials, as when pigments or filler are dispered in a plastic medium. The amount of light that is scattered depends largely on the difference in the refractive index between the two materials. For this reason pigments are most efficient as light scatters or opacifiers when their refractive index is quite different from that of the polymer with which they are to be used. The amount of light scattering also depends on the size of the scattering particles. Scattering increases with increasing particle size until the particles are about the same size as the wavelength of light, then scattering decreases with still larger particles.

The light returning to the observer's eye from the pigmented plastic is often the combination of reflected and scattered portions of the incident light, plus the balance of transmitted and absorbed light. For each pigment or combination of pigments, these portions vary at different wavelengths; as a result, the light as received by the observer is usually quite different from that emitted by the illuminating source. With many inorganic pigments the function of scattering or random reflection predominates; most organic pigments are predominantly light-absorbing.

COLOR COORDINATES

In order to study color in the limiting comparative sense, where all psychological, spatial, and environmental factors are controlled, the basic attributes common to all perceived colors can be thought of as arising from three sensory stimuli: red, green, and blue. As a result, a monochromatic light (i.e., light of a single wavelength) may produce the same visual sensation as does a mixture of several different monochromatic radiations. For example, a pure orange appears as a mixture of red and yellow. In the observer's mind, this three-variable combination is reinterpreted into another set of three sensations, hue (dominant color), value (lightness vs darkness), and chroma (saturation or purity of color). The observed differences in color between two objects can be described in terms of these three variables. They are called color coordinates and form the basis of the language system most widely used in color description and specification.

Hue

The most readily detected difference between two isolated colors, especially if the difference is small, is called hue. It is described by the commonly used color words (red, green, purple, beige, etc.) and the agreement in the common nomenclature is quite close. The total number of hues that can be distinguished may be as high as 500. Of this number, however, only four are unique. People differ widely in their observation of the four unique hues, because of color-vision differences. They are defined, however, as absolute red, yellow, green, and blue. All other possible hues are seen as mixtures of two of these, with the restriction that it is not possible to have a hue that is a mixture of blue and yellow or of red and green.

With these pairs eliminated as mixtures, the remaining pairs can be arranged in a continuous circle. The intermediate colors lie between the four absolute colors: red and yellow give orange, yellow and green give yellow-green; green and blue give cyan; and blue and red give purple. All intermediate hues can be spaced between these, and the spacing can be so-arranged that opposite colors are complementary, i.e., they produce white light if mixed properly as lights.

Saturation or Chroma

A second difference between two colors of the same or different hue is the amount of hue seen. This variable is saturation or depth of color. There is no well-defined upper limit to saturation in color, but there is a zero saturation, at which point there is no hue perception. Since hues can be arranged in a circle and saturation starts at zero and increases, the two can be combined in a circle diagram with zero at the center. To indicate a point on such a diagram a saturation scale is needed. Such a scale is usually produced by presenting equally discernible intervals of saturation, working with actual samples.

Brightness or Value

The third observable difference between two isolated colors is brightness. A color can vary from being too dim to be seen, to being too bright to be looked at. Except for difficulties in producing it, as far as perception is concerned, such a change can occur for any hue at any saturation.

The three color coordinates hue, chroma and value are sufficient to describe color, but they do not completely describe the appearance of an object. Other qualities, size, gloss, surface texture, opacity, and the colors of nearby things also affect appearance.

COLOR-ORDER SYSTEMS AND NOMENCLATURE

It has become increasingly important, in industrial operations requiring color specification, color control, and color matching, to describe colors in precise and easily comprehensible definitions. The orderly description of color is an essential part of all discussions about color, colorants, and coloring.

Collections of physical samples put together without guiding principles are common in some industries, including the plastic industry. Sample cards for vinyl-coated fabrics, plaques of cast or extruded plastic sheets, and swatches of laminating films are often used for a presentation of colors. Samples are meant to illustrate the colors available in a given product line; however, a vast reference library of color samples often accumulates over the years. This may comprise discontinued colors, samples of pilot products, results of pigment evaluation and color matching and, not the least, color samples of competitors. One serious objection to this disarray is that it makes impossible the extrapolation of intermediate color samples from the samples displayed.

Among suppliers of colorants to the plastic industry, it is common to produce a wide variety of colored samples by the systematic mixing of a relatively few highly colored pigments with one another and with white, black, and grey in a plastic medium at several concentrations. This system of interpolation between samples is helpful, but it requires the manufacturer to see samples from each supplier. A drawback of the lack of an orderly system of color description is that no common language of color has been developed, and the correlation of one sample with another is impossible without actually comparing the two samples. Efforts have been made to evolve a color-order system which does not completely depend on actual samples and which allows use of a common color language.

Perhaps the best known of all color-order systems is the Munsell system. It is both a collection of samples painted to represent equal intervals of visual perception of color difference between adjacent samples, and a system describing all possible colors in terms of the three coordinates; hue, value, and chroma.

The Munsell system of color notation consists of a hue circle made up of principle hues – red, yellow, green, blue, and purple –

and five intermediate hues. Each of the ten hues is divided into ten steps; at the center of the hue circle is the gray pole, varying in brightness from black at the bottom with a value of zero to white at the top with a value of ten. This brightness dimension is called value. From the gray pole, the colors increase in saturation as the radius of the hue circle increases, so that the most saturated colors occur at the periphery. The saturation scale is called the chroma scale. The Munsell color system is a three-dimensional model and may be visualized as a cylinder (see Fig. 2).

A representation of the system consisting of small printed chips of about 1500 colors systematically sampling Munsell spacing of color is available as the *Munsell Book of Color.* The samples are arranged in planes or pages of constant hue. On each page, samples are arranged by Munsell value in the vertical direction and by Munsell chroma in the horizontal direction. Each sample carries a Munsell notation denoting its position: 2.5 RP 3.87/4.23; wherein, the

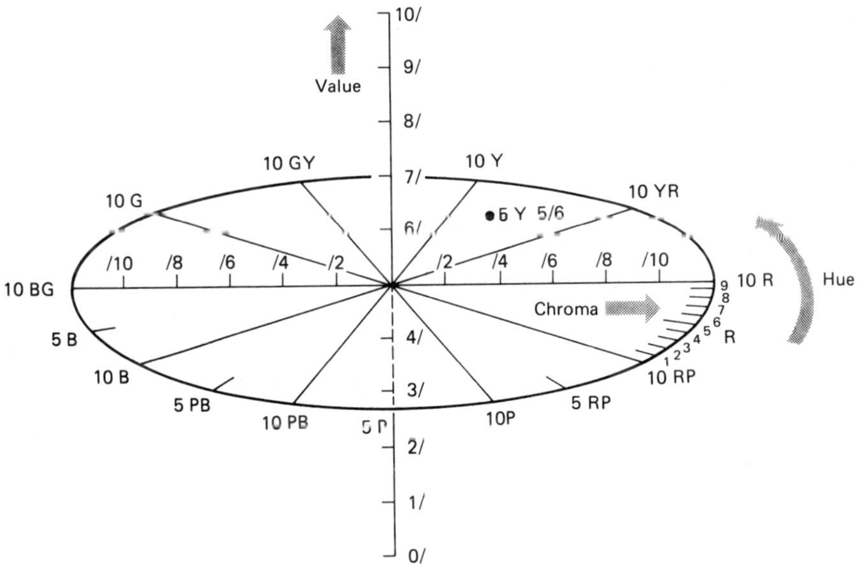

Fig. 2. In the Munsell color-order system, colors are arranged by *hue* in a hue circle, by *value* or lightness from top to bottom, and by *chroma* or saturation according to distance out from the scale of grays making up the center column.[2] Source: Billmeyer, F. W. and Saltzman, M. *Principles of Color Technology.* Interscience. New York (1966).

number-letter combination describes the hue (2.5 RP for red-purple) and the value description (3.87 for moderately dark color and chroma designation, 4.23 indicating medium saturation) are always separated by a slash.

Two outstanding features of the Munsell system contribute to its usefulness and wide acceptance. First, it conforms to equal visual perception; second, its notation is not linked to or limited by existing samples. Any conceivable color can be fitted into the system, whether it can be produced with existing colorants or not.

The Munsell system was used as the basis of the Inter-Society Color Council–National Bureau of Standards (ISCC–NBS) system for designating color names. The system was developed as an aid to standardizing the verbal description of color by names, such as very pale blue, very light blue, pale blue, light blue, brilliant blue, grayish blue, moderate blue, strong blue, vivid blue, dark blue, deep blue, etc., for different shades of blue corresponding to Munsell's hues between 9 B and 5 PB.

BIBLIOGRAPHY

1. Saltzman, Max. "Color Measurement with the Eye and Other Instruments." *Color Eng.* **I** (4), 12–18 (1963).
2. Billmeyer, F. W. and Saltzman, Max. *Principles of Color Technology.* New York: John Wiley & Sons (1966).
3. Hunter, Richard S. "Measurements of Color and Other Appearance Attributes in the Plastics Industry." *SPE J.* **23** (2), 51–58 (1967).
4. Johnston, R. M. "The Effect of Specular Reflectance on Color." SPE *Regional Technical Conference,* Philadelphia, (1972).
5. Evans, Ralph M. "The Perception of Color," pp. 43–68. In R. M. Johnston and M. Saltzman *Industrial Color Technology.* American Chemical Society, Washington, D. C. (1972).

2

Measurement of Color and Appearance

INTRODUCTION

Evaluation and measurement of color in plastics are important for several diverse, but interrelated, reasons. If color is to be specified it must be described in specific terms, possibly in numbers. It should also be measurable. In addition, measurement is necessary for color matching and control. Only with reliable measurement of the attributes of optical appearance is it possible to achieve and maintain appearance quality.

Optically, the three major classes of plastics are: transparent, translucent, and opaque. Color is important for all three classes, but the geometric manners in which color is measured vary from one optical class to another. Also changing with the optical class of specimens are the important geometric attributes which need to be evaluated and measured. Some of these geometric attributes are: gloss of reflecting surfaces, haze of transparent films and sheets, and hiding power of translucent sheets. Table 1 is a compilation of seven common types of plastic products, showing the appearance attributes that usually receive major attention.

Measurement and *evaluation* are invariably associated with the use of instruments. But the eye might also be considered an instument for measuring color and appearance. Visual examination and comparison of colored material is just as much a measurement of color

Table 1. Classification of Plastic Products by Appearance Attributes

	Clear Transparent Sheets						
	Flexible Films	Cast Sheets And Objects	Translucent Sheets and Molded Parts	Opaque Flexible Film	Opaque Cast or Laminated Sheet	Opaque Molded Objects	Extruded Shapes
Product uses	For transparent bags and packages, protective covers, windows, etc.	For windows, light closures, lenses	For lighting fixtures, luminous signs, corrugated structural components	Upholstery, covering materials	Table, counter and furniture tops, other structural uses	Plastic articles and objects for house, business, industry	Plastic pipe, strip rod and stock for machining operations
Major resins used	Cellulosic, Polyethylene, Vinyl, Polystyrene	Methacrylate, Polystyrene	Methacrylate, Polystyrene, Epoxy	Vinyl, Rubber	Melamine Methacrylate Epoxy, Rubbers Phenolic, Polystyrene	Polyethylene, Rubber, Polystyrene, Melamine, Methacrylate	Nylon, Fluorocarbon; Polyethylene, Vinyl, Rubber
Dyes and pigments used	Dyes used when color is required (Most are colorless)	Dyes used when color is required	White pigments or rough cast surfaces used to diffuse light Dyes and color pigments used to color	White and colored pigments used	White and colored pigments, dyes	White and colored pigments, dyes	White and colored pigments, dyes
Filler materials	None	None	None, except glass fiber in corrugated panels	Some extender pigment may be used	All types of fillers, limestone, wood flour, fiber glass, etc.	All types of fillers	Some inorganic fillers

Table 1. Classification of Plastic Products by Appearance Attributes (continued)

		Clear Transparent Sheets						
		Flexible Films	Cast Sheets And Objects	Translucent Sheets and Molded Parts	Opaque Flexible Film	Opaque Cast or Laminated Sheet	Opaque Molded Objects	Extruded Shapes
Important Appearance Attributes: Raw Materials	Resins	Haze Yellowness and resistance to yellowing of resins	Haze Yellowness and resistance to yellowing of resins	Yellowness and resistance to yellowing of resins	Yellowness of resins	Yellowness of resins	Yellowness of resins	Color of resins
	Other			Whiteness and diffusing power of white pigments	Color and coloring performance of pigments	Color and coloring performance of pigments	Color and coloring performance of pigments	Color of pigments
	Finished Product	Haze Transmitted-image clarity Yellowness Transmission color Gloss	Haze Transmitted-image clarity Yellowness Transmission color	Yellowness Diffuseness or "hiding" power Light transmission Color by transmission Color by reflection	Color Gloss Resistance to fading	Color Gloss Resistance to abrasion Resistance to fading	Color Gloss Resistance to fading	Color, or yellowness Diffuseness or hiding power
Is computer formulation of color applicable?		Occasionally	Occasionally	Yes	Yes	Yes	Yes	Occasionally

Source: R. S. Hunter. *SPE J.* 23 51–58 (Feb. 1967).

and appearance as are the results of measurement with an instrument. Regardless of what techniques are used, both instrumental and visual assessment of color have two things in common: 1) an object is examined, and 2) its color and appearance are assessed.

Many factors affect what we see. As described in Chapter 1, basic to the human response of color are the spectral distribution of the light source in which an object is viewed, the spectral modification made by the object being observed, and the response characteristics of the viewer. These are the variables in the viewing situation, defined by the CIE (Commission Internationale l'Eclairage) on which the whole technology of color measurement and evaluation is based. But other aspects of the human viewing situation are also very important to the final color perception. Some of these affect both instrumental and human viewing. The complexity of human perception affects such variables as the state of the viewer's adaptation, the nature of the area surrounding the viewing scene, etc. Materials characteristics which affect both human and instrumental evaluations are the presence of fluorescence, the degree of homogeneity, the texture of the surface, the internal structure, the relative transparency, etc. The method, or the combination of methods, which is most appropriate to the sample characteristics must therefore be ascertained.

Color and appearance measurement, whether visual or instrumental, is nothing more than a special technique of analysis. As such, it shares with all other analytical techniques the fundamental problem of obtaining a representative sample. Whether a sample truly represents the material being examined is the first question to be asked before undertaking an evaluation. The question of selecting a sample for inspection and color measurement includes the further step of converting the sample into a form suitable for inspection. While there may be cases where a finished article already exists in a form suitable for either visual or instrumental examination after proper sampling, most materials require that the sample be handled in some special way. For example, a sample of translucent film must be folded so that a standard number of layers is presented for examination.

The conversion of a sample of colored material into a form suitable for examination requires a standardized procedure which is both repeatable and reproducible. Only then can the color of a batch of material be discussed and compared.

Converting a colored object into a form suitable for inspection is especially important where color measurement is to be done by instruments.

Although visual and instrumental methods of color evaluation are not fundamentally different, for convenience in our discussion of color measurement we will separate wholly visual methods and methods employing some instruments. Both types of measuring techniques rest heavily on experience.

VISUAL ASSESSMENT OF COLOR AND APPEARANCE

The visual examination and evaluation of color is the most commonly used technique in the plastics industry. There are many standard methods for making this type of measurement (e.g., ASTM D 1535, D 1729). However, no instrument can perceive small color differences as quickly and accurately as can the human eye. Unfortunately, individual observers do not always agree on these differences – and many people have poor color memory. The gist of all efforts in visual color evaluation is to counter these shortcomings by providing standards, fixing tolerance limits, and specifying standard viewing conditions. In the simplest form, the two essential elements of visual color evaluation are 1) the viewing of both the sample and a standard by an experienced observer, and 2) the use of a predetermined light source at a fixed distance.

Disagreement between manufacturer and customer may result if the source of light is not specified, or if the slight differences in observers' perceptions are not taken into account. In many cases a single light source is not adequate for proper inspection, and two or more standardized sources may be agreed upon. With the proliferation of types of light, especially those with illumination quite different from daylight or tungsten lamps, it has become almost essential to use at least two standardized sources of illumination for the examination.

Standardization of the observer is, of course, not possible. Differences among observers, all with normal color vision according to the usual tests, can lead to at least as great a variation in the judgment of color as can the use of different light sources. Unless observer differences are recognized by prior agreement, no amount of standardization of sample and light sources can lead to satisfactory results.

The use of a single standard against which a sample is compared leads to immediate difficulties if sample and standard are not identical because each item represents only one point in a three-dimensional world of color. If the sample and standard do not match we must know both *how* and by *how much* they differ. This brings us to quantitative and analytical judgments at which the eye is not very adept. This limitation of the human eye can be compensated for by use of more than one standard for comparisons. In fact, visual examination with the aid of one or more limit standards in addition to the target standard is becoming quite common. The number of limit standards, as well as their degree of difference from the target, depends upon the level of color tolerance involved. Where necessary, fairly elaborate systems have been evolved. In such systems, limit standards are set up in three directions from the target standard – the high and low limits for lightness and darkness (value), hue and saturation (chroma). The use of a complete set of limit standards combined with standardized illuminants provides a very satisfactory procedure for plant control work.

It is important to remember that there is frequently a great difference between the conditions under which colored materials are examined and those under which they are used. The use of color-coded wire and cable is an excellent example. Here it is required that two different-colored wires always look *different* when seen under field conditions. Color standards which can comply with this requirement are quite different from those needed to control the color of objects which must look exactly *alike* under viewing conditions which permit the slightest color diffference to be perceived. A related example is the plastic components of a car exterior which have to match the bodycolor paint in all viewing conditions. Limit standards must be adapted to both the tolerance level involved and the purpose for which color control is being sought.

INSTRUMENTAL METHODS OF COLOR MEASUREMENT

In ordinary usage we speak of color measurement, however, in reality we do not *measure* color with an instrument. Rather, we *describe* values which correlate with visual evaluations of color in idealized and standardized terms, derived from a measurement of the light reflected, transmitted, or emitted by a specimen. Color is what we see; the color of an object is the result of the physical modification of light by colorants in the object as observed by the human eye and interpreted in the brain.

In general, the purpose of color measurement is to provide numerical correlation with visual evaluation; the degree to which this can be accomplished depends on the conditions of both the measurement and the visual evaluation. The objective for instrumental measurement and specification of color is to provide both qualitative and quantitative information on the nature and size of small color differences. Such measurement can also provide a quantitative memory for color, something the eye and brain can do only to a limited degree. As of this writing these objectives have only been partially realized, although great effort is being expended to bring the area of color measurement and appearance evaluation fully under instrumental techniques. Because of constant progress in the field, a complete description of the state-of-art of instrumental color and appearance measurement is neither possible nor desirable. The literature on this subject is varied and abundant; the interested reader is referred to the Bibliography.

In replacing the eye with an instrument for color measurement, one is replacing a detector which is very sensitive to qualitative differences with one which is less sensitive but has a much greater ability to measure and remember in a quantitative way. In order to obtain numbers, and the same numbers every time, we use instruments; thereby, sacrificing the human observer's ability to look at a sample in nearly any light and report its hue, lightness, saturation, and many other appearance aspects.

Sometimes for expedience, sometimes for convenience, and often because of the varied requirements of different color-measurement problems, many different types of color-measuring instruments are in use. The two major types are the *spectrophotometers* and the *tristimulus colorimeters.* In one respect the function of these two instrument types is similar, for after data reduction both furnish the tristimulus values of the sample. Actually, however, the spectrophotometer provides much more information on the spectral reflectance curve of the sample. There are many important areas of color technology which depend on this information, including the detection and control of metamerism and the formulation of mixtures of colorants having specified colors. In other areas, the tristimulus values alone suffice, and either type of instrument can be used. Because of their simplicity and lower cost, tristimulus colorimeters are used more often for color-difference measurement, but spectrophotometers can do this job just as well.

The instrumental color measurement has its roots in the CIE color-order system. The CIE system of color description enables every possible light stimulus to be defined by three numbers, known as the tristimulus values, *X, Y, Z.* These numbers can be regarded as the axes of a three-dimensional euclidean space in which every perceivable color can be located, not only the color of objects such as pigmented plastics but also the colors produced by monochromatic light, the purest color sensations perceivable. Just as color invariably involves three components (a source of light, an illuminated object, and an observer), the CIE system requires and provides both a standard light source and a standard observer to complete the description of the color of objects.

The physical quantities important in specifying color are the spectral power distribution of the light source (CIE standard source),

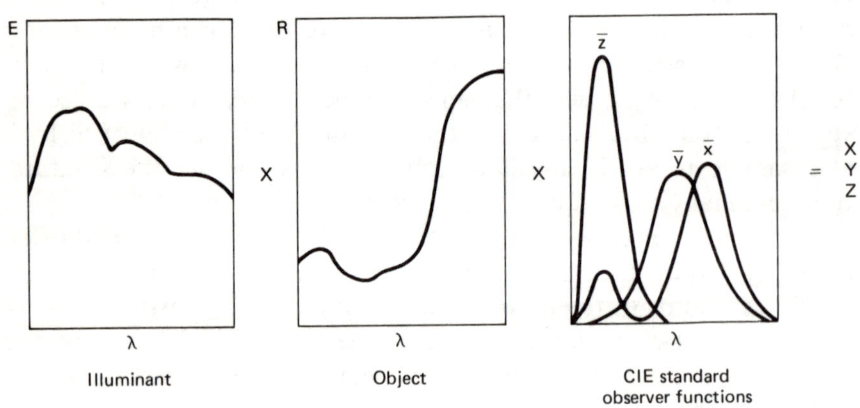

Fig. 3. Schematics of the CIE coordinate system. Source: Osmer, D. *SPE ANTEC* (1977).

the spectral reflectance of the sample, and the spectral response of the eye in the form of color-matching functions (those of CIE standard observer, *X, Y, Z*). Because the eye performs an integrating rather than analytical function, the three components must be multiplied and integrated over the visible wavelength region to calculate the color coordinates (tristimulus values) which will describe the color of the sample (Fig. 3).

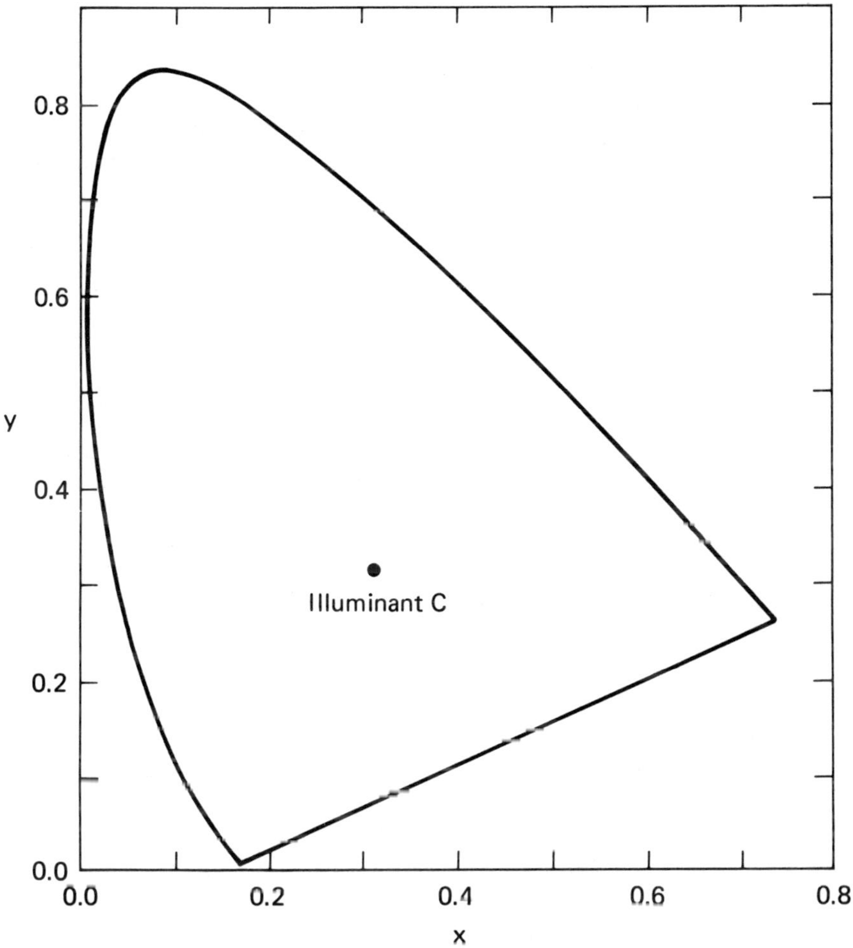

Fig. 4. Chromaticity diagram with two dimensions of color space. Source: Osmer, D. *SPE ANTEC* (1977).

Chromaticity coordinates for evaluating hue and saturation are calculated by

$$x = \frac{X}{x + y + z}$$

$$y = \frac{Y}{x + y + z}$$

These coordinates then define a chromaticity diagram which contains two dimensions of color space (Fig. 4).

The third dimension, lightness, is included by adding the lightness scale, Y, perpendicular to the chromaticity diagram (Fig. 5). The

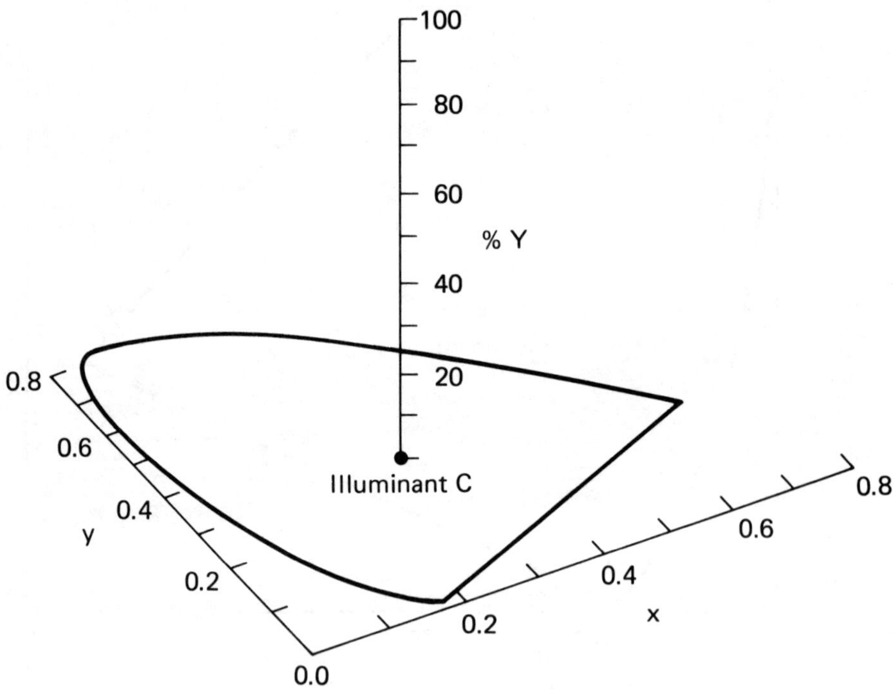

Fig. 5. Chromaticity diagram with hue, saturation, and lightness coordinates. Source: Osmer, D. *SPE ANTEC* (1977).

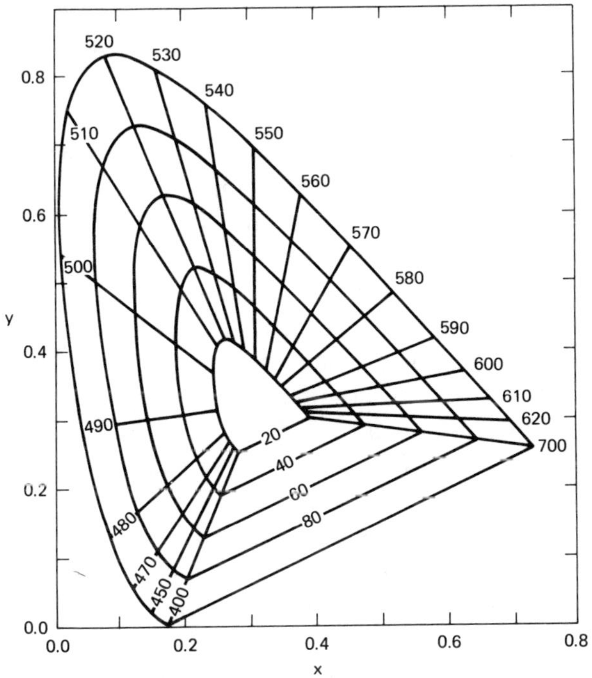

Fig. 6. CIE color space in terms of dominant wavelengths. Source: Osmer, D. *SPE ANTEC* (1977).

third dimension must always be taken into account, because two colors might be the same with respect to hue and saturation but different in lightness. Therefore, it is possible to have identical chromaticity coordinates and yet not be at the same lightness level. The following example illustrates this point:

	Sample 1	Sample 2
X	40.00	8.00
Y	50.00	10.00
Z	60.00	12.00
x	0.2667	0.2667
y	0.3333	0.3333

In addition to describing color by tristimulus values (X, Y, Z) or chromaticity coordinates (x, y) and lightness (Y), the CIE offers a third possibility, dominant wavelength, purity, and lightness. Dominant wavelength and purity roughly correlate with the visual aspects of hue and saturation. The CIE diagram shown in Fig. 6 demonstrates dominant wavelength and purity.

The dominant wavelength coordinate provides a hue description which correlates with the visible spectrum. The nonspectral colors, e.g., purples, are described by complementary (generally green) dominant wavelengths. Purity is the relative distance from the illuminant point along the dominant, and is useful for describing saturation. Figure 7 shows the names of the various sections of color space.

The dominant wavelength and purity diagram is based on color additivity. The lines of dominant wavelength all intersect at the illuminant point (Illuminant C). By mixing various amounts of pure monochromatic light of a certain wavelength, and the illuminant, the color would in fact proceed along the dominant wavelength line corresponding to the wavelength of the monochromatic light. The distance traveled from the illuminant point (purity) would correspond to the relative amounts of monochromatic and illuminant light.

Using a dominant wavelength and purity coordinate system, the chromaticity diagram becomes a powerful tool for analysing coloristic behavior. If mixtures of colored lights were plotted, straight lines would result. Because colorants selectively absorb and scatter light, their characteristic behavior is not necessarily a straight line. This is because the resultant color of a combination of different lights is based on an additive color mixture, while the resultant color is based on a subtractive colorant mixture.

If, for example, two different yellow samples are plotted, a determination can be made as to which color is greener or redder than the other, and which color is more saturated or pure. By expanding this and plotting each colorant at several tint levels, one can learn about the coloristic characteristics. It must be remembered that these diagrams contain only two dimensions. Because we are looking down at the chromaticity diagram, differences in lightness are not detected. Therefore, even if two colorant paths intersect, they do not necessarily match at these concentrations. At the crossover point, both would have the same degree of chroma and purity; yet, the degree of lightness might be different.

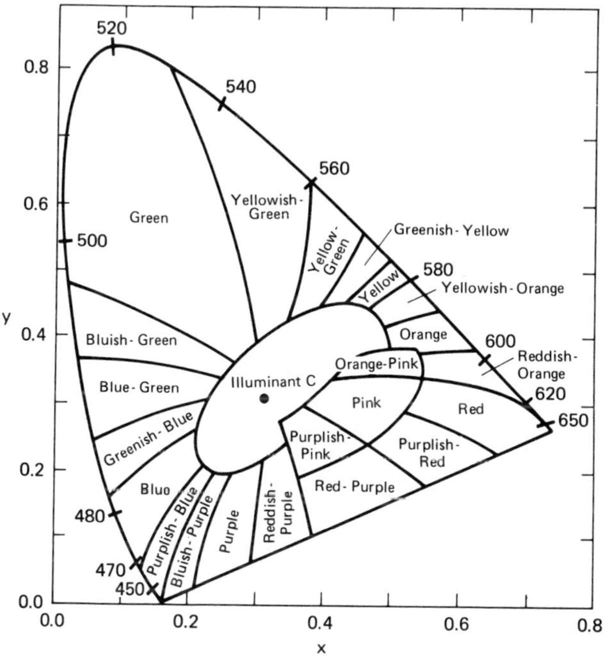

Fig. 7. Chromaticity diagram in terms of various color names. Source: Osmer, D. *SPE ANTEC* (1977).

A word of caution is needed here. To describe such a complex phenomenon as color (itself only a small part of the total appearance of a sample) by the specification of only the three CIE tristimulus values, is a gross oversimplification. The great value of the science of tristimulus colorimetry lies in its usefulness in describing the relationship between colors. Thus, there is no absolute color description – the notations determined by the computational procedures depend on the method by which the spectral data are obtained, on the computational procedures used, and on the relationships of these factors to the nature of the samples being compared.

BIBLIOGRAPHY

1. Billmeyer, F. W., Jr. and Saltzman, M. *Principles of Color Technology*. New York: *Interscience* (1966).
2. Le Grand, Y. *Light, Colour and Vision*, 2nd Ed. London: Chapman & Hall Ltd. (1968).

3. Hal-Curtis. "Evaluation and Description of Metallic Colors." SPE Annual Technical Conference **XI**, Boston (1965).
4. Williams, J. M. "Color Measurement of Metal Surfaces." *J. Opt. Soc. Am.* **51** (June 1961).
5. Johnston, R. M. "Color Measuring Instruments: A Guide to their Selection." *Color and Appearance* **1** (2) (Sept./Oct. 1971).
6. Johnston, R. M. "Measurement of the Transmittance and Reflectance of Translucent Plastics." SPE Regional Technical Conference, Cherry Hill, New Jersey (Sept. 24, 1973).
7. Johnston, R. M. and Saltzman, Max. (ed.) *Industrial Color Technology.* American Chemical Society, Washington, D. C. (1972).
8. Bugash, R. S. "A C Technical Offers Quantuum Advantage in Operation of Appearance Measuring Instruments." SPE Regional Technical Conference, Cherry Hill, New Jersey (Sept. 24, 1973).
9. Rosenthal, R. D. "Instrument-to-Instrument Repeatability: the Key to Widespread Use of Color Measurement Instruments in the Plastics Industry." SPE Regional Technical Conference, Cherry Hill, New Jersey (Sept. 24,1973).
10. Coppock, W. A. "The Requirements of a Spectrophotometer/Colorimeter in a Plastics Coloring Operation." SPE Regional Technical Conference, Cherry Hill, New Jersey (Sept. 24, 1973).
11. Danby, R. E., Jr. "Color, Color Measurement and Colorant Formulation in the Textile Industry." *JAATCC* **88/47** (Sept. 1973).
12. Johnston, R. M. "Applications of Color Difference Formulas." *J. Paint Technol.* **42** (542) pp. 145–150, (Mar. 1970).
13. Venable, W., Jr. "Specification and Measurement of Optical Properties of Translucent Materials." SPE Regional Technical Conference, Cherry Hill, New Jersey (Sept. 24, 1973).
14. Stanziola, R. "Current Practices in Measuring Plastics, Including the Many Uses for Color Measurement." SPE Regional Technical Conference, Cherry Hill, New Jersey (Sept. 24, 1973).
15. Conner, E. T. "The Promise of Improved Measurement and Application." SPE Regional Technical Conference, Cherry Hill, New Jersey (Sept. 24, 1973).
16. Saltzman, M. "Common Pitfalls in Measurement of Plastics." SPE Regional Technical Conference, Cherry Hill, New Jersey (Sept. 24, 1973).
17. Hunter, R. S. "Measurements of Color and Other Appearance Attributes in the Plastics Industry." *SPE J* **23** (2), 51–58 (1967).
18. Holtzen, D. W. "Spectrophotometric Matching of Opaque Plaques." SPE Annual Technical Conference, (1975).
19. Osmer, D. "Using Chromaticity Diagrams for Pigment Selection." *Plastics Eng.*, p. 38 (Sept. 1977).
20. Nanfeld, R. E. "A Method of Measuring Color Development of Color Concentrates." SPE Annual Technical Conference (1977).

21. Hunter, R. S. "Progress in Achieving Reproducibility in Worldwide Appearance Measurements of Plastics." SPE Annual Technical Conference (1978).
22. McCamy, C. S. "Recent Developments in Appearance Measurements." SPE Annual Technical Conference (1978).
23. Osmer, D. "Opacity Parameter." SPE Annual Technical Conference (1978).

3
Colorants

INTRODUCTION

The substances that modify the perceived color of objects, or impart color to otherwise colorless objects, are collectively called *colorants.* They produce color by selectively absorbing and reflecting specific areas of light energy from wave bands which constitute white light. The colorants are traditionally divided into two broad classes, dyes and pigments.

Dyes are synthetic organic chemicals soluble in most solvents and in the medium which they color. They are characterized by good transparency, high tinctorial strength, and low specific gravity. *Pigments* are intensely colored solids, insoluble in common solvents. In contrast to dyes, pigments exhibit no affinity for the materials in which they are used. They produce color by existing as discrete particles in the medium. Because of their insolubility, pigments must be uniformly and finely dispersed in the plastic by mechanical means to provide optimum properties. Pigments provide color but they also diffuse light; therefore, they impart translucent or opaque color to plastics.

Even though there are definite differences between dyes and pigments, precise definition of a dye or a pigment is not possible. If we take the criterion of solubility, there are several pigments which are to some extent soluble in a variety of solvents and plastics, and there are some dyes which are soluble in common solvents but not in plastics. If we take the criterion of transparency, we may find that many pigments can be so well dispersed in plastics that the resulting colored resin is completely transparent. The same is true if we consider specific gravity or any other properties of colorants. Although the distinction between the two groups of colorants is open to considerable interpretation, we will continue discussing the available colorants as belonging to one group or the other on the basis of their predominant characteristics.

While dyes are invariably of organic origin, pigments can be inorganic or organic in structure. The two groups of pigments may differ widely in their properties from each other; the individual pigments within a particular group having much in common. The inorganic pigments are insoluble in common solvents, yet the majority of the organic pigments exhibit some degree of solubility in a few organic solvents. Inorganic pigments are more temperature-resistant than the organic pigments. Color intensity, on the other hand, is generally better in organic than inorganic pigments.

Although not all coloring matters are suitable for coloration of plastics, most of the inorganic pigments and numerous organic colorants are at the disposal of the plastic colorist. Solvent dyes are suitable for many transparent colors, and there are 30 or more types of pigment available for translucent or opaque colors. Other appearance-modifying substances (see Chapter 4) are also sometimes used, alone or in combination with dyes and pigments. In this Chapter, we will discuss those substances which impart color to the plastic and thus truly fit the definition of a colorant: 1) inorganic pigments, 2) organic pigments, and 3) dyes. As plastics are very diverse in chemical nature, it is not surprising that not all types of colorants and appearance-modifying agents are suitable for use with every plastic. It is necessary, therefore, for the plastic colorist to know the general properties of each of these categories and to appreciate that some of them are subject to considerable subdivision, also that numerous brands of the same colorant may be available in the market.

INORGANIC PIGMENTS

Inorganic pigments can be either naturally occurring or synthetic. They are usually metallic oxides, sulfides, lead salts, cadmium and mercury compounds, and carbon blacks. As a group they are insoluble in common solvents as well as in the plastic materials. They are in general superior to organic pigments in heat and light stability. With the exception of lead salts, chrome yellows, and molybdate oranges, they are stable at all plastic compounding temperatures. They are characterized by high specific gravities of 3.5 to 5.0.

Inorganic pigments are relatively easy to disperse in plastics. When used in high concentrations they yield fairly bright colors, although not as bright as those possible with organic pigments and dyes. Tinctorially, the inorganics are weaker than organics. Inorganic pigments are normally used in plastics where opaque or translucent colors are required. Table 2 lists some of the important inorganic pigments which are suitable to varying degrees for the coloration of plastics. Some of the most important ones are described in the following pages.

Inorganic pigments have retained their importance as colorants for the plastic industry in an increasingly competitive and sophisticated market. This has not been due to the discovery of new chromophores or to the development of revolutionary new processes, but mainly to some outstanding properties of the well-established pigments and to the steady improvement and modification to combat pigment weaknesses. Inorganic colorants constitute an important section of the pigment market. Their output is similar in value to that of organic pigments and much higher in quantity (see Table 3).

Oxides

Iron Oxides

Iron oxides are both synthetic and naturally occurring. Natural iron oxides offer a very inexpensive range of subtle color. However, they are of poor texture and quality. They are therefore used in the plastics industry only for such applications as vinyl floor tiles, where no stringent demands on quality and color tolerances are made. Synthetic iron oxides are commercially the most important class of colored inorganic pigments used in the plastic industry. They range in hue from yellows to black.

Iron oxide reds and browns are used more extensively in plastics than blacks and yellows. They have good heat stability, good

Table 2. Performance properties of organic pigments in plastics.

Common Name	Chemical Composition	C.I. Designation	Hue	Brightness	Tinctorial Strength	Transparency	Acid Resistance – Dilute	Alkali Resistance – Dilute	Heat Resistance	Lightfastness	Weathering Resistance	Migration Resistance	Toxicity
Cadmium yellow	Mixture of Cds and ZnS	Yellow 37	Yellow	E	P	F	F	E	E	E	F	E	T
Cadmium orange	CdSe	Orange 20	Orange	F	P	P	E	E	E	E	F	E	T
Mercadmium pigments	Mixed Cds–Hgs	Orange 23	Orange to maroon	F	P	P	F	E	E	E	F	E	T
Cadmium red	Mixed CdS–CdSe	Red 108	Red	F	P	P	F	E	E	E	F	E	T
Red iron oxide	Fe_2O_3	Red 101, 102	Red	F	P	P	F	E	E	E	E	E	C
Brown iron oxide	Fe_2O_3	Brown 6, 7	Buff, brown	F	P	P	P	E	F	E	E	E	C
Yellow iron oxide	$Fe_2O_3 \cdot H_2O$	Yellow 42, 43	Yellow	F	P	P	P	E	F	E	E	E	C
Black iron oxide	Fe_3O_4	Black 11	Black	P	P	P	F	E	F	E	E	E	C
Chromium oxide, anhyd.	Cr_2O_3	Green 17	Dull green	P	P	P	E	E	E	E	E	E	L
Cobalt aluminum blue	$CoO \cdot Al_2O_3$	Blue 28	Redish blue to turquoise	F	P	P	E	E	E	E	E	E	A
Chrome cobalt alumina	Cr_2O_3 $CoO \cdot Al_2O_3$			F	P	P	E	E	E	E	E	E	A

Table 2. Performance properties of organic pigments in plastics. (continued)

Common Name	Chemical Composition	C.I. Designation	Hue	Brightness	Tinctorial Strength	Transparency	Acid Resistance — Dilute	Alkali Resistance — Dilute	Heat Resistance	Lightfastness	Weathering Resistance	Migration Resistance	Toxicity
Chrome yellow	$PbCrO_4$, $PbSO_4$	Yellow 34	Turquoise to lemon yellow	F	P	P	F	F	P	F	F	E	L
Chrome orange	$(PbCrO_4)$, $(PbO)_y$	Orange 21	Orange	F	P	P	F	F	P/F	F	F	E	L
Chrome green	Mixture of iron blue and chrome yellow	Green 15	Green	F	P	P	F	P	P	F	E	E	L
Molybdate orange	$PbCrO_4$, $PbSO_4$, $PbMoO_4$	Red 104	Orange	F	F	P	F	P/F	F	F	F	E	T
Zinc chromate	Basic zinc chromates	Yellow 36	Yellow	F	F	P	E	P	F	F	P	E	L
Hititanium yellow	$NiO \cdot 7IO_2$	Yellow 53	Yellow	F	P	P	E	E	E	E	E	E	M
Titanium brown	Titanate		Brown	F	P	P	E	E	E	E	E	E	—
Titanium blue	Titanate		Blue	F	P	P	E	E	E	E	E	E	—
Titanium green	Titanate		Green	F	P	P	E	E	E	E	E	E	—
Ultramarines	Alumina citicate Complexes	Blue 29, Violet 5	Blue, violet, pink	F	P	F	P	F	E	E	F	E	L
Ceramic yellows	Sb, Ti, Cr oxide		Yellow/orange	F	P	P	E	E	E	E	E	E	T

Table 3. Consumption of inorganic colorants by the plastic industry. (in metric tons)

	1974	1975	1976	1977
Inorganics				
Titanium dioxide	94,000	66,000	90,000	92,000
Iron oxides	3700	2660	3400	3530
Cadmiums	2750	2200	2600	2680
Chrome yellows	2900	2070	2700	2750
Molybdate oranges	1900	1300	1700	1750
Others	1700	1140	1500	1510
Total	**106,950**	**75,370**	**101,900**	**104,220**

Source: *Modern Plastics,* Sept. 1977

resistance to alkali and dilute acid, and good light and weathering fastness. Generally speaking, the synthetic red and brown iron oxides are recommended for all plastics, except those which require processing temperatures above 260°C. Yellows and black iron oxides are somewhat less resistant to heat. They are generally not suitable for processing temperatures above 150°C. In using these pigments with PVC, heat stabilizers composed of lead salts should be avoided.

Iron oxides have traditionally been known for their color opacity. However, iron oxide pigments have recently been introduced that are transparent *and* metallic. In these pigments, advantage has been taken of the relationship between the particle size of the pigment and the refractive index of the medium. A transparent and metallic red iron oxide flake pigment has been prepared by steam hydrolysis of ferric chloride in a salt melt. Low opacity in iron oxides is also achieved by laking iron oxides with aluminum hydroxides, which is itself transparent in oleoresinous media. These pigments are used exclusively in low-temperature processes. Fluid-energy milling of iron oxide has led to improvements in properties, including easy dispersiblity, high gloss, and bright hues.

Chrome oxides

Anhydrous chromium oxide green is the only green inorganic pigment used in the coloration of plastics. It is made by calcining either sodium or potassium bichromate under reducing conditions. Although it is dull, poor in tinctorial strength, and low in opacity, it exhibits outstanding heat, light, and chemical stability. Hydrated chromium oxide is not widely used for coloring plastics because of its low resistance to heat. Chromium oxide is an abrasive material, and should not be used in such applications as coloration of fibers where abrasion of spinnerette holes and tow-cutting blades would be detrimental to process economies.

Red Lead Oxides

These pigments, reddish orange in hue, are made by heating metallic lead to form litharge (PbO), and then by further heating form Pb_3O_4. These pigments are not widely used in plastics since they are sensitive to reduction to metallic lead.

Sulfides

Cadmium and mercury sulfides are the most important members of this family of pigments. They range from bright yellow to maroon in color. Cadmium yellows and oranges are made by coprecipitation of cadmium sulfide with a small amount of zinc sulfide. Cadmium sulfide is orange; small amounts of zinc sulfide shifts the hue to lemon yellow. These pigments have excellent heat stability, cleanliness of hue, and resistance to light and to alkalies. Except for their toxicity and rather poor weathering characteristics, cadmium pigments would be considered acceptable for most plastics applications.

Cadmium reds are prepared by coprecipitation of cadmium sulfide with cadmium selenide. Depending on the ratio of sulfide to selenide, the shade varies from yellow red to maroon. They exhibit excellent resistance to heat, light, and chemicals, and good resistance to weathering. In the latter characteristic they differ from the yellow and orange members of the family. Deep red to maroon cadmium pigments are more resistant to fading than other shades when exposed to sunlight in the presence of moisture. These pigments are recommended for use in all polymers.

Mercadmium pigments are based on a mixed crystal of cadmium sulfide and mercury sulfide. They are a relatively new class of maroon to orange pigments, in which the shade is controlled by increasing or decreasing the concentration of mercury sulfide instead of cadmium selenide. They are lower in cost than cadmium pigments and, except for poorer heat stability, display the same performance characteristics. Mercadmium pigments are recommended for use in all polymers where heat resistance is not critical. Like cadmium pigments, mercadmium pigments are considered toxic and are not recommended for use in food-packaging materials.

Despite current concern over the toxicity of pigments based on heavy metals, the consumption of cadmium pigments has continued to rise. Products with a low-level of acid-extractable cadmium, which meet the European specification for colorants in plastics in contact with foods, are now available.

Chromates, Molybdates, Aluminates and Titanates

Chrome yellows and chrome oranges

Chromium yellow and orange pigments are lead chromates that are precipitated from a lead salt solution by the addition of a mixed bichromate and sulfate solution. An acid solution gives yellow, and an alkaline solution gives an orange color. Both are fairly bright, ranging in shade from pale greenish (lemon) to deep orange. They have good-to-excellent resistance to migration, are easily ground, and undergo little or no color change on being subjected to temperatures as high as 180°C. They darken on exposure to light, although light-stable types have appeared on the market in recent years. Chrome yellows are sensitive to alkalies and chrome oranges are sensitive to acids. As with other lead-containing pigments, their use in plastics intended for children's toys should be avoided because of possible toxicity.

Zinc yellows

Zinc yellow pigments are complex zinc chromates. They exhibit fair resistance to heat, poor resistance to weathering, poor resistance to alkalies, fair-to-good resistance to dilute acids, excellent resistance to solvent bleeding, and fair hiding power. In general, they are recommended for use in plastics that will not be exposed to sunlight and weathering. They are used most in cellulosics and general-purpose polystyrene.

Molybdate oranges

Molybdate oranges are precipitations of lead chromate ($PbCrO_4$), lead sulfate ($PbSO_4$), and lead molybdate ($PbMoO_4$) in varying proportions. They are characterized by excellent cleanliness of hue, high strength, good light fastness, good heat stability, resistance to dilute acids, and sensitivity to alkalies. Because of their outstanding strength and cleanliness of hue, they are used in blends with both organic and inorganic pigments to produce lighter shades of red, which have lower cost and better hiding power than pure toners of similar colors. In general, they are recommended for use in all polymers and applications in which chrome yellows and oranges can be used.

Cobalt aluminates

Cobalt aluminate pigments are essentially crystalline combinations of cobalt and aluminum oxides. They vary in shade from a violet reddish blue to turquoise depending on increased or decreased alumina content or a slight addition of chrome or silica. Because of their excellent resistance to heat, light, and chemicals, and their ease of dispersibility, cobalt aluminate pigments are widely used in plastics as a tinting coloring. Their use as prime colorants is precluded by their high price compared to phthalocyanine blues.

Nickel titanate

This pigment, a yellow colorant, is a mixed crystal of nickel and titanium oxides. Because of its exceptional light, heat, and chemical resistance, it is used extensively where chemical resistance and stability to high processing temperatures are required. Its low hiding and tinting strength prevent more general use. A cheaper modification of nickel titanate yellow containing barium titanate is also available and is considered suitable for plastics.

Ultramarine blue

This pigment is an aluminosilicate complex containing sulfur. It is made by a two-stage calcination of china clay, sulfur, and sodium sulfate or carbonate with coke, pitch, or rosin as a reducing agent. It is characterized by cleanliness of hue, reddish blue masstone, excellent resistance to heat, light, and weathering and to most chemicals, and good dispersibility. Its major disadvantages are a sensitivity to acids, low hiding power, and somewhat weak tinting strength. To increase its hiding power, this pigment is normally used in combination with small amounts of opaque white pigments. Because of its excellent resistance to heat and its other favorable properties, ultramarine blue is used extensively as a blue colorant in plastics. It is recommended for use in all plastics that do not come in contact with acids. Excellent stabilization of the resin must be achieved before it can be used with polyvinyl chloride.

Cobalt blue

A light blue colorant, is normally considered part of the ultramarine blue family. It is a mixture of ultramarine blue and zinc oxide, and it exhibits the same performance characteristics as ultramarine blue.

Ceramic Pigments

During the past few years considerable attention has been given to the development of ceramic pigments, generally used in ceramics and glazes. There is some evidence that the plastics industry is being compelled to consider this type of pigment, despite its high cost and low tinting strength, for the coloration of plastics and long-life surface coatings in an effort to meet the increasing demands for permanence and heat stability.

Carbon Blacks

The world production of carbon black exceeds that of any other pigment, even titanium dioxide, but only 5% is used as a pigment; the bulk is used in the rubber industry, particularly in tires.

Carbon black is produced by two different methods; incomplete combustion of natural gas (channel black), or liquid hydrocarbons (lamp black and furnace black). The channel black process is fast becoming obsolete in favor of furnace black.

The carbon content of carbon black varies from 90 to 99%. The remainder consists of organic material, mainly hydrocarbon. The rubber grades are generally 98% carbon and are of larger particle size than are the pigmentary blacks, which usually have had an oxidation aftertreatment to modify their surface properties.

The physical and chemical properties of carbon black, and the properties' relationship to performance, have been extensively studied, as indicated in the Bibliography.

The channel blacks marketed for pigmentation are often characterized by *regular color, medium color* or *high color.* This terminology has also been preserved in newly developed furnace black, although this tells the plastic colorist little about the tinting strength of the grade in question. More important is the structure of a particular grade of carbon black. *Structure* is a measure of the degree of irregularity in the shape of primary aggregates. Highly branched aggregates create more void space, which accounts for higher oil absorption and lower density. The structure is measured by dibutylthalate absorption, nitrogen absorption, and specific resistivity of conditioned carbon black. These values for furnace black used as a pigment generally fall in the range of 200 to 350 cc/100g, 200 to 300 sq. meters/g and 0.2 to 0.9 ohm, respectively. All other analytical properties being equal, low-structure furnace blacks have higher tinting strength than do high-structure samples. However, high-structure grades disperse more easily compared to low-structure grades.

ORGANIC PIGMENTS

All organic pigments are synthetic and are benzenoid in nature. Several types of organic pigments have been developed and quite a few of them are well suited to the coloring of plastics.

Compared to inorganics, organic pigments are generally lower in specific gravity, higher in surface area, more transparent, stronger in color, higher in price (although not necessarily more expensive to use), and clearer in shades. Unfortunately, some fade badly on exposure to light, deteriorate on exposure to chemicals, cannot withstand high processing temperatures, and bleed in solvents. The pigment chemist, by preparing pigments of much larger molecular size, is overcoming some of these defects. This research for large molecules stems from the elucidation, some 10 years ago, of the structure of the phthalocyanine pigments and from the realization that their excellent properties were due in part to their large size. At present, almost all of these modern high-grade organic pigments are too expensive to displace inorganic pigments, although for some uses they may be ultimately cheaper; also, in some applications, their merits may outweigh their higher costs.

Table 4. Performance properties of organic pigments in plastics.

Common Name	C.I. Designation or Chemical Type	Hue	Brightness	Tinctorial Strength	Transparency	Acid Resistance	Alkali Resistance	Heat Resistance	Lightfastness	Weathering Resistance	Migration Resistance	Toxicity	Dispersibility
Hansa	Monoazo	Yellow, orange, red	E	E	E	F	F	F	F	P	P	A	F
Diarylide	Diazo	Yellow, orange	E	E	E	E	E	F	F	F	A		
Nickel azo	Chelated azo	Greenish yellow	F	E	E								
Condensation azo	Condensation azo	Yellow, orange, red	E	E	E								
Quinacridones	Quinac	Gold, orange, red	F	E	E								
Flavan throne	Vat Anthraquinone	Yellow	E	E	E								
Anthrapyrimidine	Vat Anthraquinone	Yellow	E	E	E								
Orange RK	Vat Anthraquinone	Orange	E	E	F								
Orange CR	Vat Anthraquinone	Orange	E	E	F								
Isoindolinone	Isoindolinone	Yellow, orange, red	E	E	E								
Para, Parachlor, Tolnidnie, Arylide	Monoazo	Orange, red, maroons	E	E	E								
Pyrazolone	Diazo	Orange, red	E	E	E								
Lithol	Precipitated azos	Red, maroon	E	E	F								
Red 2B, lake	Precipitated azos	Red	E	E	F								
Lake red C	Precipitated azos	Red	E	E	F								
Bon	Precipitated azos	Red	E	E	F								
Lithol rubin	Precipitated azos	Red	E	E	F								
Thioindigo	Vat	Red, maroon, violet	E	E	E								
Anthraquinone red	Vat	Red	E	E	E								
Anthraquinone yellow	Vat	Yellow	E	E	E								
Perylene, perinone	Vat	Red	E	E	E								
Carbarole dioxazine	Oxazine	Violet	E	E	E								
Phthalocyanines	Phathalocynines	Blue, green	E	E	E								
Indanthrone	Vat	Blue	E	F	E								

Organic pigments vary widely in chemical structure, thereby accounting to a large degree for their wide range of colors and other properties. Products of similar composition can be expected to behave alike; hence, classification according to chemical type, irrespective of color, is useful in performance prediction. The commercially available organic pigments that can be considered for the coloration of plastics (listed in approximate order of the date of their introduction) are:

Phthalocyanines
Vats
- Anthraquinone
- Thioindigoid
- Perinone and perylene

Azo
Dioxazine
Quinacridone
Isoindolinone

These pigments are briefly described in the following pages. The performance properties of some of the organic pigments are given in Table 4.

In spite of their generally higher cost and limited heat stability, organic pigments have been used increasingly over the past few years (see Table 5). The Table consumption of carbon blacks, sometimes classified as organic pigments, is given for comparison. Although poundage volume of organics is considerably lower than the volume of inorganics, they are fairly equal in dollar volume.

Table 5. Consumption of organic pigments by the plastics industry. (in metric tons)

	1974	1975	1976	1977
Organics				
Carbon black	27,000	19,400	26,000	26,900
Phthalocyanine blues	1700	1300	1600	1660
Phthalocyanine greens	1000	700	900	930
Organic reds	1200	890	1100	1140
Organic yellows	200	160	190	200
Others	540	390	500	520
Total	31,640	22,840	30,290	31,350

Source: *Modern Plastics*, Sept. 1977

Phthalocyanine Pigments

The phthalocyanines come the closest to representing the ultimate in pigment performance in the plastic industry. They are the technical leaders in blues and greens. Excellent fastness to heat, light, chemicals, and solvents is combined with high-tinting strength and transparency, low-use cost, and intense pure shades. The hue ranges from reddish blue to green. The blue exists in at least two crystal forms, a red shade alpha form and the more stable green shade of beta form. The greens are derived by chlorination of the blue; yellow greens are made by substituting bromine for chlorine. Bright green shades containing both chlorine and bromine are also available.

Most of the phthalocyanines available today are copper based, although metal-free products and some based on other metals are also available. Copper phthalocyanine blues have a tendency to drift to greener or lighter in systems where slight solubility exists at a high temperature (e.g., polystyrene and nylon).

Phthalocyanine pigments are not easy to disperse. They were, therefore, originally offered to the plastics industry in reduced form, i.e., in the form of intimate mixtures with inert materials that disperse more easily than the parent pigments. Even today, some of the concentrated forms of the green phthalocyanine pigments are too difficult to disperse adequately in some plastics, and the predispersed pigments must be utilized to achieve an acceptable quality.

Vat Pigments

Vat dyes have been known to the textile industry for their excellent brilliance of shade and light stability. Vat pigments are chemically related to these dyes: they contain two or more carbonyl (C=0) groups in their molecules. Vat pigments include those compounds based on thioindigo, anthraquinone, perinone, and perylene structures. These pigments cover a wide range of hues and are generally of high strength, transparency and, intense coloration. Fastness properties vary greatly, even within each chemical group.

Anthraquinone pigments have acceptable lightfastness when used in most plastics, but the heat stability of the reds and yellows limit their use to plastics that are processed at 250° to 280°C. Heat stability of blue and violet anthraquinone pigments is, however, satisfactory. The anthraquinones are particularly susceptible to loss through chemical reduction. They also have a tendency to bleed and migrate.

As a group the thioindigos have the poorest heat stability. They are, therefore, not suitable for use in plastics that are processed at high temperatures. Like anthraquinones, they have a tendency to bleed and migrate. Their lightfastness is only fair. The C.I. Pigment Red 88 and C.I. Pigment 87 can be used in plasticized PVC, but if good light stability is required, relatively high concentrations of pigments must be used.

Perylene and perinone pigments (e.g., C.I. Vat Red 29 and C.I. Vat Orange 27) are outstanding for their high-heat stability and good lightfastness. They can satisfactorily be used in coloring such thermoplastics as polypropylene and nylon.

Azo Pigments

The azo pigment class is the largest of the organics in terms of both volume of consumption and variety of products available. Their colors are almost entirely in the yellow, orange, and red hues. They are characterized chemically by the presence of an azo chromophore (–N=N–) in the molecule which connects the aromatic groups of varying complexity. Azo pigments can be of monoazo type (containing one chromophore) or polyazo type (containing two or more chromophores). Poor heat stability and strong bleeding tendencies severely restrict the use of monoazo pigments in plastics. The polyazos are generally more heat and bleed resistant, but the improvement is not sufficient to permit their widespread use in thermoplastics. However, excellent chemical resistance makes the polyazos specially suitable for elastomeric and thermosetting systems where strong curing and oxidizing agents are often used. None of the polyazo colorants is suitably lightfast in low concentration for long-term outdoor use.

The metallized azos are monoazos of varying chemical structure that are precipitated from a metal salt, metal complex, or laked on an absorptive substrate. The metal salts and lakes are almost entirely reds and maroons. Heat stability varies from poor to good within the group, with Lake Red C and Red 2B pigments favored for high-temperature applications. A number of pigments in this group drift dark and blue when hot, but return to their original color when cooled. Lightfastness is poor in low concentrations, but Red 2B types have found acceptance for noncritical outdoor applications in dark shakes. The Red 2B pigments in particular tend to plate out with other ingredients in improperly formulated vinyls. Also, metal used in the manufacture of precipitated azos can be injurious to some polymers. One azo of importance is the nickel azo yellow. In contrast to other precipitated azos, this one offers excellent lightfastness in all concentrations, along with good heat stability and bleed resistance. Although chemical resistance is generally good, the relatively weak chelate bond makes the pigment subject to demetallization, which in turn degrades its other properties.

Condensation azo yellows and reds are high-molecular weight diazo pigments that are structurally similar to conventional azo colorants, but their higher molecular weight gives them generally better bleed resistance than diazo dyestuffs. They are high-strength, bright colors of relatively high cost. Tint lightfastness varies within the group, but it is decidedly better than in most other azo pigments. These pigments have sufficient heat stability to permit their use in plastics that are processed at medium temperatures. However, to ensure good light stability, high concentrations must be used. The C.I. Pigment Yellow 93 and 94, and C.I. Pigment Red 142 and 166 in plasticized PVC with a barium-cadmium stabilizer system perform satisfactorily. They can also be used in PVC containing a lead-based stabilizer, but once more a high concentration is necessary to achieve good light stability.

Dioxazine Pigments

This class of pigment was the outcome of exploratory work on polycyclic compounds persued by color chemists searching for improved pigments in the vat dye field. The first pigment of this class, the violet carbazoledioxazine (C.I. Pigment Violet 23) developed by Farbwerke Hoechst, is the only one firmly established so far. Redder varieties have recently been produced, although it appears that not all their migratory problems have been overcome. However, C.I. Pigment Violet 23 has excellent stability to light and satisfactory, though limited, heat stability, and so can be used in acrylic sheets and extrusion grades of low-density polyethylene. It is also widely used for shading greenish phthalocyanine blues to red. This is one of the best violet pigments on the market, even with its limited heat stability.

Quinacridone Pigments

The quinacridones come the closest to duplicating the excellent performance of phthalocyanines in the orange, red, and violet portions of the spectrum. Good lightfastness in all concentrations, excellent bleed and chemical resistance, and adequate heat stability for processing in most resins has led to wide adoption of quinacridones in spite of their high cost. They perform in polystyrene and acrylic resins best in the lower temperature range because of their limited heat stability. Because they are slightly soluble in nylons, quinacridones are generally not recommended for use in these resins.

A number of the quinacridone products are difficult to disperse. It is interesting to note that C.I. Pigment Violet 19 can be obtained in two hues, an illustration of the way a radical change can be wrought in a pigment by altering its physical characteristics (particle size and surface texture), while retaining its chemical structure.

Isoindolinone Pigments

Isoindolinones constitute a new class of high-performance organic colorants. At present there are four types (two yellows, an orange, and a red), all marketed under the trademark IRGAZINE (Ciba-Geigy). Because of their relative newness, little information is available on how extensively they are used in plastics. They are recommended for use in most polymers, except nylon, that do not require processing temperatures above 40°C for long exposures and 290°C for 1 to 2 min. In nylon, they tend to change in color to a yellow shade, possibly as a result of a solution effect in the polymer.

DYES

Dyes are chemical compounds which absorb certain wavelengths of light while transmitting others and scattering none. Thus, unlike pigments, they are transparent. Dyes may be water soluble or oil soluble. Those used for coloring plastics are generally of the latter type. They are characterized by bright and intense colors. Dyes are usually most suitable for rigid transparent plastics such as polystyrenes, acrylics, polycarbonates, polyesters, and unplasticized PVC and cellulose acetate. They have poorer resistance to light and heat than do pigments and have a tendency to bleed or migrate in some flexible plastics, particularly in polyethylene, plasticized PVC, and ethylene-vinyl acetate copolymers.

Dyes are the only means of coloration available for perfectly transparent plastics. Bluish dyes are used with a number of transparent resins when it is necessary to counteract the natural yellowish tint of the raw resin. In spite of their limited use, the consumption of dyes in plastics is considerable (see Table 6).

As is the case with organic pigments, fastness properties of dyes are dependent of their chemical structure and their end use. Yellow, orange, and red shades are generally based on a diazo structure while violet, blue, and green shades contain anthraquinone or phthalocyanine chromophores. Having relatively small molecules, the diazo oil-soluble dyes are tinctorially strong; they exhibit moderately bright hues. Anthraquinone-based dyes are somewhat less soluble

Table 6. Consumption of dyes by the plastics industry. (in metric tons)

Dyes	1974	1975	1976	1977
Nigrosines[a]	1500	1050	1430	1500
Oil-soluble types	690	520	620	650
Anthraquinones	210	160	200	210
Metal complex types	170	130	150	155
Others	130	90	110	115
Total	**2700**	**1950**	**2510**	**2630**

Source: *Modern Plastics*, Sept. 1977

than are the diazo types, but they exhibit better stability to heat and light.

Dyes are organic and therefore can be oxidized; for this reason processing conditions greatly influence the selection of dyestuffs. It should be noted also that, because most organic coloring agents are antioxidants, the colorist must content not only with bleaching but also with the degree to which the curing of a plastic will be inhibited.

One must exercise caution in selecting only those dyestuffs which are known to be free of diluents. Diluents have an adverse effect on the working properties of the dye and tend to opacify where transparency is desired. They also appear to have an adverse effect on electrical properties.

The color industry has kept abreast of the rapid growth of rigid thermoplastics and is able to offer much assistance and advice in the application of dyestuffs. Various data have been developed which enable colorists to make recommendations for specific plastics. Table 7 lists some commercially available dyes, which can be used as colorants for rigid thermoplastics.

Solvent-soluble dyes yield bright transparent shades in general-purpose polystyrene. However, since these dyes have poor bleed resistance in lacquer, they are not recommended for general-purpose polystyrene where lacquer decorations are used. The same dyes can be used in modified polystyrene, but their performance depends on the extent of modification.

Colorants suitable for general-purpose polystyrene are generally applicable to acrylic molding compounds and cellulosics. For the

Table 7. Solvent-soluble dyes for general-purpose polystyrene, acrylic molding compounds and cellulosics.

Designation	Other Applications
Solvent Violet 13	Rigid PVC, polycarbonate
Solvent Violet 14	Rigid PVC
Solvent Green 3	Rigid PVC, polycarbonate
Solvent Red 111	Rigid PVC
Solvent Yellow 33	Polycarbonate
Solvent Red 24	

Source: *Modern Plastics*, Sept. 1977

latter, their performance depends largely on the amount of plasticizer used; thus, dye concentrations have to be adjusted to prevent migration and exudation.

Unplasticized (rigid) PVC can be colored with four of the dyes shown in Table 7. Relatively low concentrations yield pleasing shades with good light stability and do not impair the vinyl compound's clarity. Blue, yellow, and orange dyes are also commercially available for rigid PVC.

Three of the dyes listed in the table are suitable for polycarbonate molding components. Red and orange dyes, not listed here, are also available for polycarbonates.

Available Forms

The forms in which colorants are available to the plastic industry are:

1. Dry colors
   a) single pigments
   b) pigment blends
2. Treated dry colors
   a) single pigments
   b) pigment blends
3. Pigment dispersions
   a) resinuous dispersions
   b) solvent dispersions
   c) aqueous dispersions
   d) plasticizer's dispersions or pastes
4. Color concentrates
   a) single pigments and dyes
   b) pigment blends
   c) custom concentrates
5. Custom colored molding chip or powder

An essential step in the use of colorants in plastics is the dispersion of the pigment in the medium. To facilitate better dispersion, several kinds of surface treatments with surface-active agents are used. These pigments are called treated or flushed pigments. Actually, items 1 and 2 in the above list can be combined into one group, because treated dry colors are in fact extensions of regular dry colors.

The term *surface treatments,* refers to small amounts of added materials, in the range of 0.5 to 2%, or slightly more. When larger amounts of added materials (e.g., inert pigments) are used to aid dispersion, the general term is *reduced pigment.* For these products, the added materials can be from 5 to 80%, and they usually reduce the color strength proportionately to the amount used. One type of reduced pigment is a *lake,* which causes some confusion. The classic definition of a lake is a dyestuff precipitated on an inert pigment, such as barium sulfate or aluminum hydrate, which then serves as a color pigment. Since a reduced pigment made by physically mixing a 100% color pigment and a specified amount of inert pigment is similar in appearance to a lake, some manufacturers use the term interchangeably with reduced pigment.

Pigment *dispersions* are an intimate mixture of finely dispersed pigments in a suitable *carrier.* The carrier is generally a resinous material, such as low molecular-weight polyethylene, waxes, ethylene-vinyl acetate copolymers, etc. It can also be an active solvent, especially for oil-soluble dyes and organic pigments. Aqueous dispersions are not common, but they are available for special uses such as in polyurethane foam and viscose rayon applications. Pigment pastes and plasticizer dispersions are of interest to the vinyl industry.

Because of dispersion problems, the color concentrate industry has found it necessary to produce predispersed concentrates on a custom basis. It can supply any shade, in any degree of strength, made from any combination of pigments, in any type of plastic – provided, of course, that the order is sufficient to justify the expenses. While the cost is higher if color concentrates rather than the basic pigment, are used, the dispersion is very often better.

Cost is, of course, an important consideration in selecting any of the forms of pigments. It has already been mentioned that dry color forms are, under most conditions, probably the least expensive to employ for coloring. Single-pigment concentrates and custom-colored concentrates follow in that order, and the colored molding powders are the most expensive. The choice of the form of pigment is also governed by the criticality of the process and the tolerances in quality control. In several applications, the use of dry colors is

Table 8. Growth pattern of color systems. (in million lb)

Coloring System	1971	1976	1981 (estimated)
Pellet color concentrate	300.0	640.0	100.0
Multifunctional concentrate containing color	200.0	20.0	100.0
Dry color	40.0	140.0	160.0
Paste dispersion	2.0	50.0	60.0
Proprietary encapsulate	–	4.0	8.0
Direct-feed liquid	–	8.0	40.0
Other	–	–	8.0

Source: *Modern Plastics*, Sept. 1977.

not possible and predispersed color concentrates have to be employed. These problems are discussed elsewhere in this book in detail. Table 8 summarizes the growth pattern of various color systems in the plastic industry, and Table 9 gives a breakdown of colorant consumption by major plastics groups.

Table 9. Colorants consumed by major plastics groups.

Market	Resin Containing Colorants, 1000 metric tons 1976	1977	Compound, % '76	'77	Dry Color, % '76	'77	Pellet Concentrate, % '76	'77	Total Colorants Consumed, metric tons, '76	'77
Low density PE										
Blow molding	26	28	4	3	12	12	84	85	440	470
Coatings	35	36	10	9	N	N	90	91	4070[d]	4180
Film	375	390	6	5	N	N	94	95	12,400	12,450
Sheet and profiles	8	8	10	10	N	N	90	90	500	500
Injection molding	350	350	5	4	40	41	55	55	3000	3000
Wire and cable	180	220	10	8	15	15	75	77	2250	2750
Others	250	285	10	8	15	16	75	76	2300	2600
Total	**1224**	**1317**							**24,960**	**25,950**
High density PE										
Blow molding	490	500	5	4	10	10	85	86	5300	5470
Injection molding	270	272	10	9	36	37	54	54	2800	2830
Pipe and profiles	100	105	50	40	25	30	25	30	2000	2100
Wire and cable	30	40	54	50	18	18	28	32	270	350
Others	140	160	34	32	34	35	32	33	1400	1600
Total	**1030**	**1077**							**11,770**	**12,350**

Table 9. Colorants consumed by major plastics groups. (continued)

Market	Resin Containing Colorants, 1000 metric tons		Com-pound, %		Dry Color, %		Pellet Con-centrate, %		Total Colorants Consumed, metric tons,	
	1976	1977	'76	'77	'76	'77	'76	'77	'76	'77
Polyvinyl chloride										
Calendering	500	500	N	N	46	46	54	54	11,500	11,500
Coatings	250	230	6	6	44	42	50	52	5320	4700
Extrusion	600	670	10	9	48	48	42	43	9100	10,050
Molding	200	210	11	10	58	58	31	32	2100	2200
Others	180	190	2	2	74	74	24	24	2000	2010
Total	**1730**	**1800**							**30,020**	**30,460**
Polystyrene										
Injection molding	550	570	10	8	40	42	50	50	6800	6930
Sheet extrusion	200	210	30	25	20	20	55	55	6470	6790
Others	400	400	25	20	52	54	23	26	4000	4000
Total	**1150**	**1180**							**17,270**	**17,720**
Polypropylene										
Injection molding	500	520	10	9	34	35	56	56	6120	6400
multi/ monofilaments	250	240	2	2	N	N	98	98	2600	2500
Sheet extrusion	9	9	80	80	N	N	20	20	230	230
Others	75	80	32	31	12	13	56	56	1300	1370
Total	**834**	**849**							**10,250**	**10,500**
ABS										
Injection molding	140	154	40	39	12	13	48	48	2600	2860
Pipe extrusion	70	85	55	50	5	6	40	44	1460	1750
Sheet extrusion	80	84	48	46	5	5	47	49	1600	1680
Others	30	24	48	48	5	5	47	47	1350	1080
Total	**320**	**347**							**7010**	**7370**
Grand total	**6288**	**6570**							**101,280**	**104,350**

Source: *Modern Plastics,* Sept. 1977.

BIBLIOGRAPHY

1. Nadolski, F. J. "Colorants for Plastics." *SPE J.* p. 1076 (Sept. 1965).
2. Smith, F. M. "Phthalocyanine Pigments." *J. Oil Colour Chem. Assoc.* **49** 614–630 (1966).
3. Papillo, P. T. "Mass Coloring of Plastics." *Mod. Plast.* **45**, p. 131 (Dec. 1967).
4. Ehrich, F. F. "Pigments, Organic." *Kirk-Othman Encyclopedia of Chemical Technology.* **15**, 2nd Ed., pp. 555–589 New York: John Wiley (1968).
5. "Pigments" in *Encyclopedia of Polymer Science and Technology* **10**, pp. 157–188. New York: *Interscience* (1968).
6. Schwarz, H. F. "Phthalocyanine Blue Pigments for Plastics." SPE Annual Technical Conference **XIV**. New York (1968).
7. Moilliot, J. A. and Plant, D. A. "Surface Treatment of Pigments." *JOCCA* **52**, 289–308 (1969).
8. Bartolo, H. F. "Quinacridone Pigments in Plastics." SPE Annual Technical Conference **XV**. Chicago (1969).
9. Balley, E. J. G. "The Coloration of Plastics." *J. Soc. Dyers Colour.* pp. 571–578 (Dec. 1969).
10. Thornley, D. G. C. "Pigments, Plastics, Progress." *JCDC.* pp. 13–19 (Jan. 1970).
11. Reeve, T. B. "Organic Pigments for Plastics." *J. Macromol. Sci. Rev. Poly. Tech.* **D1** (2) 217–235 (1971).
12. Mess, W. H. and Garett, M. D. "The Dispersion of Carbon Black." *J. Oil Colour Chem. Assoc.* **54**, 24–83 (1971).
13. Moon, W. R. "Organic Pigments – Available Forms." *Pigm. Resin Technol.*, p. 24 (Sept. 1972).
14. Smucklar, J. H. "Furnace Black Replacements for Channel Blacks in Plastics Systems." SPE Regional Technical Conference, Philadelphia (1972).
15. Barnes, P. J. "Proposed Classification for Pigments." *Pigm. Resin Technol.*, pp. 13–20 (Sept. 1972).
16. Taylor, P. J. "Colored Inorganic Pigments." (142 Ref.) *Rev. Prog. Color.* **4** (1973).
17. Wiese, V. K. "New Aspects in Dry Coloring of Polyolefins with Organic Pigments." SPE Annual Technical Conference (1975).
18. Reeve, T. B. "Colorants." *Modern Plastics Encyclopedia* (1976).
19. Richter, J. "Azo Pigments with Improved Fastness Properties and Higher Covering Power for the Coloration of Plastics." SPE Annual Technical Conference (1978).

4 Special Effect Pigments

INTRODUCTION

Just as pigments are used to impart color to plastic, so are various groups of compounds used in the plastic industry to modify other appearance characteristics. One of these noncoloring groups of "pigments" is the delustering or whitening pigments, whose sole function is to diffuse light without altering its special composition. They are widely used in making plastic products that are both white and opaque. Other noncoloring pigments are used in the plastics industry to impart brilliance or glow. These are colorless organic compounds collectively called "optical brighteners" or "brightening agents." Still other color- or appearance-modifying agents, loosely grouped as special-effect pigments, are daylight fluorescent dyes, metallic-look pigments, and nacreous pigments. Appearance can also be modified by polymeric materials. One important example is the marble effect created by a mixture of optically incompatible plastics.

OPACIFIERS

Chapter 1 describes how light may be transmitted, absorbed, and scattered by a pigment-polymer system. If the light is both selectively absorbed and scattered, we perceive an opaque or translucent color. However, if the scattering of light occurs without selective absorption, the perceived color is white and opaque. The function of the colorants is to selectively absorb light while at the same time scattering or transmitting it. However, there are some pigments available to the plastics industry whose main function is to scatter light *without* selective absorption. These pigments are known as delustrants, opacifiers, or dulling agents. They may be used alone for white opaque plastics or combined with colorants to impart a high degree of opacity.

Scattering of light by a pigment-plastic system is a function of the difference of refractive indices of the pigment and the plastic. It also depends on the particle size of the pigment relative to the wavelength of light. The refractive indices of plastics are relatively small. Thus, the scattering power of a pigment is effectively proportional to its refractive index. Table 10 lists the refractive indices of some white pigments considered as opacifiers for plastics. As the refractive index increases, the scattering power of the pigment-polymer system (the *opacifying capability* or *covering power* of the pigment) increases.

Calcium carbonate (chalk) and calcium sulfate are naturally occurring minerals and quite low in price. Because of their low-refractive indices, however, they are not commonly used as opacifiers in the plastics industry. The only exception is where they are used as fillers or delustrants in such noncritical applications as vinyl flooring. Barium carbonate and barium sulfate are not commonly available at reasonable prices and are not often used by the plastics

Table 10. Refractive indices and hiding power of some white pigments.

Opacifier	Refractive Index	Hiding Power
Chalk	1.45	increases ↓
Calcium carbonate	1.57	
Calcium sulfate	1.59	
Barium carbonate	1.60	
Barium sulfate	1.64	
Zinc oxide	2.00	
Lead white	2.01	
Antimony oxide	2.19	
Zinc sulfide	2.37	
Titanium dioxide (anatase)	2.55	
Rutile	2.70	

industry. Lead white has been used in the past in plastics, because it also performs as a heat stabilizer. However, it greys quickly in the atmosphere, and its toxic effects forbid its use in many applications. Zinc sulfide, mainly as lithopone (30 to 60% ZnS with $BaSO_4$) offers a good white, but with the availability of titanium oxide pigments its use has been abandoned.

The white pigment used most, by far, in the plastics industry, is TiO_2. Zinc oxide and antimony oxide are also used in significant amounts, because each has some specific properties possessed by no other white pigment. The former is an effective UV absorber and is noted for preventing the yellowing of white plastics. Being basic, it readily forms zinc soaps which can lead to crocking, but the same zinc soaps apparently inhibit mold growth on plastic. Antimony trioxide is a very pure white, but it is relatively expensive. It is used mainly because of the fire-retarding properties it acquires when used in conjunction with halogenated compounds.

A high refractive index does not by itself assure maximum whiteness, brightness, and opacity in plastics. Opacity is developed from the reflectance and refraction of light at the pigment-binder interfaces: the greater the number of interfaces that can be formed from a given quantity of pigment, the greater the dulling power or opacity that will be produced. Of course, there is a limit to the process of increasing opacity through size reduction.

The ability of a particle to scatter or diffuse light of a particular wavelength is a function of its particle size in relation to the incident wavelength. The most effective particle diameter is approximately one-half the wavelength of the incident light. If particle size reduction is advanced to less than half the shortest wavelength of visible light, the particle will be unable to effectively reflect or refract light; the light waves will simply go around the particle. Thus, for optimum efficiency, opacifiers are manufactured to produce a maximum number of particles that fall in the 0.2- to 0.35-μ range, which is exactly one-half the 400- to 700-nm range of the spectrum of visible light. A white pigment with a 0.2-μ average particle size should produce a bluer reflectance than a 0.3-μ pigment since its scattering is more efficient at the 400-nm wavelength. Its total reflectance will also be greater due to the larger number of pigment particles present per pound of pigment.

Titanium dioxide (TiO_2) is the most important white pigment used in plastics, and therefore deserves detailed treatment in this discussion. Currently, more than four hundred grades of titanium dioxide pigments are available from over 30 suppliers. They differ in pigment particle size, particle size distribution, inorganic or organic surface treatments, and ease in dispersibility. The difficult area confronting the color compounder is the choice of the titanium oxide most suitable for the application at hand.

Titanium oxide pigments are produced either from anatase or from rutile grade TiO_2. The two varieties differ in their crystal structure. Where blue-tone whiteness is the major criterion in whites, the anatase grades are often used. The lower abrasion characteristics of anatase favor its application as a delustering agent for synthetic fibers and cut-rubber threads. However, due to its lower refractive index, its tinting strength or hiding power is about 30% lower than the rutile grade TiO_2. The weathering resistance of anatase is also lower than that of rutile. In an exposure test it was shown that acrylic sheets, pigmented with anatase variety of TiO_2, began to crack and flake after exposure for 12 to 18 months to atmosphere in the United Kingdom, whereas little change in appearance was shown by sheets pigmented with rutile, even after exposure for 10 years. In view of its lower tinting strength and inferior weatherability, the use of anatase TiO_2 in plastics is rather limited.

Rutile TiO_2 specialty pigments of fine particle size and uniform particle size distribution have been developed for the plastics industry. A much more important development is connected with their surface treatments. Almost from the start titanium dioxide pigments have been modified by coating with various inorganic compounds to improve their weathering and dispersibility properties, and more recently a further stage of posttreating with organic compounds has been introduced to achieve even better performance. It is in this sphere of activity that the plastics technologist hopes to reap the most benefit. Special grades of this type have already been produced for use in rigid PVC to reduce, and possibly eliminate, the tendency of the pigment to plate-out. This phenomenon is not a simple one and is not associated with the pigment alone; the type of filler used also plays an important part. However, as stated, specially treated titanium dioxide pigments have been introduced which show a marked improvement over the older types of pigments.

The fine particle size 0.2-μ rutile TiO_2 pigments are most effective in scattering and diffusing 400 nm wavelength of light, producing blue tone with high reflectance. Conversely, the larger particle-size grades, 0.35 μ are most efficient at 700 nm with lower reflectance and less blue tone. Just like reflected tones, the transmission tone effect in translucent white sheeting or ceiling panels is vitally affected by the particle size of TiO_2. The fine particle-size grades, while reflecting blue, transmit the least blue tone. The larger particle-size pigments provide the bluest or whitest tonal transmission.

Particle size distribution is also a vital factor. Pigments characterized by the same average particle size need not produce similar opticals since their distribution curves may be totally dissimilar, markedly differing the relative amounts of fine and coarse particles. Pigment-particle classification is a major production tool to provide the degree of strength designed into a pigment grade.

Fine particle-size TiO_2 seems to be the evident choice, as far as opticals are concerned, for all plastic compositions. However, plastisols is the one significant exception. At both high and low-TiO_2 concentration, the larger particle-size grades produce significantly higher strength than do the finer grades – a performance that should be anticipated at high but not at low-concentration level.

OPTICAL BRIGHTENERS

The optical brighteners, sometimes called fluorescent whitening agents (FWA), are a special class of essentially colorless fluorescent organic substances which absorb UV radiation of about 3000 Å and emit this energy as visible radiation below 5500 Å. They are thus related to daylight fluorescent pigments as opacifiers are related to normal pigments. They are used mostly in the production of synthetic fibers, outdoor signs, and decorative panelling. Optical brighteners are useful in correcting slight discoloration produced by some stabilizers and other additives in plastics, to enhance the whiteness of a plastic surface and to brighten the color of pigmented articles.

Two main types of optical brighteners are commercially available. One group, which resembles common water soluble dyes is used extensively for the optical brightening of textiles. The other group is insoluble in water but may be soluble in organic solvents; it is most suitable for use in plastics. Some of the optical brighteners in this group are quite stable at high processing temperatures and can be used for most thermoplastics.

The efficiency of optical brighteners is dependent on several factors: the intensity of the fluorescent emittance, the gradient of the emittance curve in the UV range, the quality of dispersion or solution in the polymer, and the concentration. The first two factors are a function of chemical structure of the compound. A few structures known to be of practical significance are shown in Table 11.

Table 11. Chemical structures of some optical brightening agents.

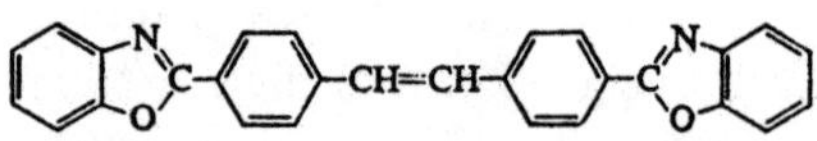

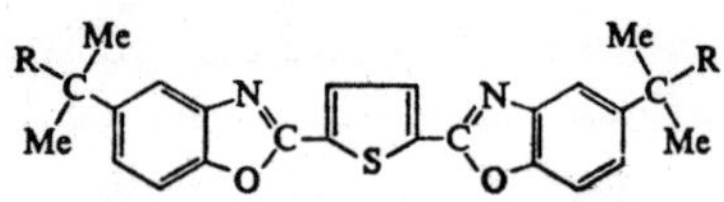

I; R = Me   II; R = Ph

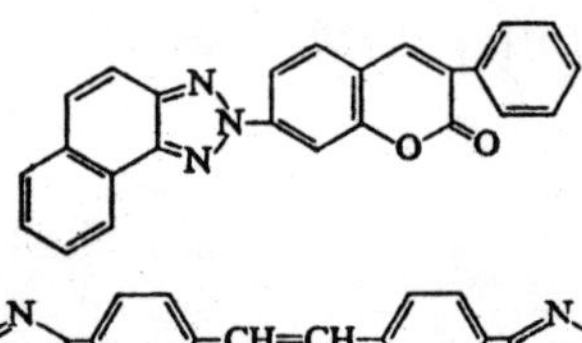

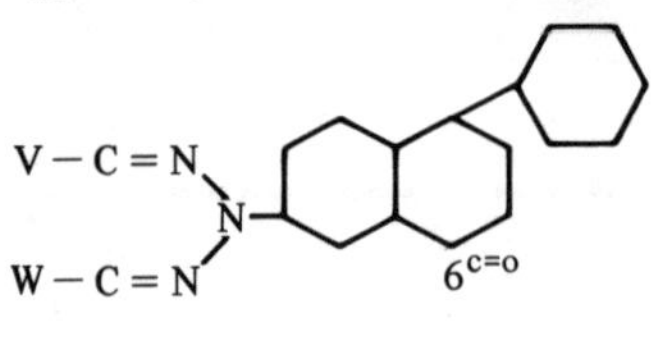

V & W = $C_{1\text{-}15}$

Molecular dispersion or complete solution of optical brighteners is essential for achieving high efficiency in optical brightening. This can be seen from the fact that very low concentrations of oil-soluble brighteners would impart a significant optical effect in nylon and polyester fibers when applied to them from water dispersion, but fail to show any effect when mixed into the melt prior to its being spun into fibers. The cause is that absorption from water dispersion dissolves the optical brightener in the polymer matrix in a molecular form, but mixing the brightener into the polymer melts at best disperses it in the form of aggregates. This difficulty can be avoided by "dyeing" a suitable white polymer with the optical brightening agent and using the "dyed" chip or powder polymer as a concentrate. Another way to circumvent the difficulties of molecular dispersion is to "dye" a cross-linked, clear, hard resin with the brightener, and then mill this mass to fine particle-size "pigment." Development work is being done in this direction, but so far no such commercial product is available.

The concentration at which optical brighteners should be used depend on their chemical structure, the degree of dispersion, and the nature of plasticizer or other additives present in the plastic formulation: the normal concentration range is 0.02 to 0.2%.

Since the mode of action of an optical brightener involves the transformation of UV light into visible blue or violet light, the components that will have the greatest influence on it are those which ab sorb UV light, namely, white pigments and UV stabilizers.

The brightening action of FWA's depends on the nature of other pigments and UV absorbers present in the formulation. White pigments vary in the extent to which they absorb UV light. That part of the UV range over which most optical brighteners are activated is also the range over which, for example, anatase and rutile types of titanium dioxide are particularly absorbent, and zinc sulfide is somewhat less absorbent. These differences are most noticeable in brightener-containing white-pigmented molded articles. When the same concentration of optical brightener is used, the most intense whiteness is generally attained with zinc sulfide and the least intense whiteness with rutile-ty] e TiO_2. With the latter pigment, it is possible that under certain conditions the whiteness will not be appreciably improved. When an improvement does occur, a limiting value of the white effect is soon reached, as the concentration of optically whitening agent is increased, e.g., at 2.0% anatase and 0.2% optical brightener and at 5% rutile and 0.05% optical brightener in PVC formulations.

Figure 8 shows UV absorption over the range 300 to 390 nm for two UV stabilizers widely used in light stabilization of plastics. It also shows UV absorption for a bis- (benzoxazolye) derivative optically brightening agent. Where the absorption ranges of UV absorbers and optical brighteners overlap, the optical brightener in the surface layer is deprived of UV light; the result is less whiteness than would be possible in formulations that did not contain any strong UV absorber. Because of this limitation the use of optical brighteners is recommended only in those formulations which contain no UV absorbers.

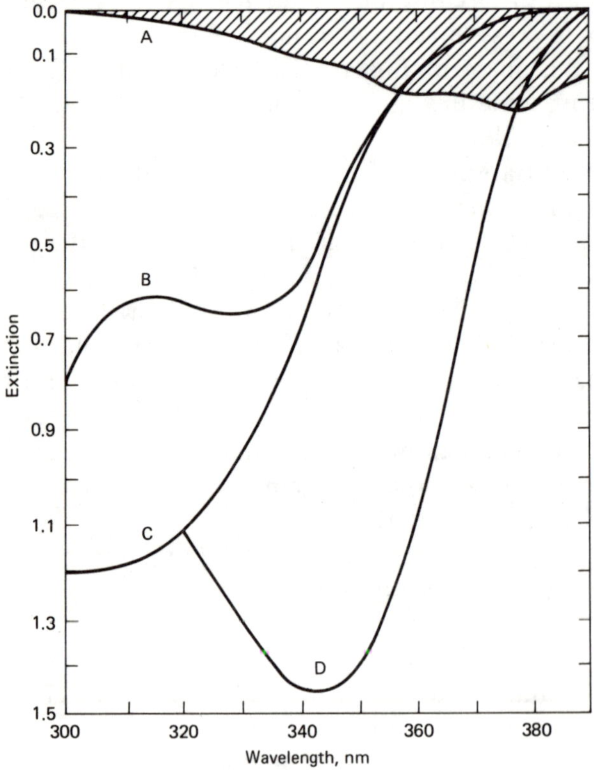

Fig. 8. UV absorption of OBA and UV absorbers. A = 0.002% OBA; B = 0.002% UV stabilizer, benzophenone type; C = 0.002% Oxalic acid-diarylamide; D = 0.002% UV stabilizer, benzotriazole type. Source: Preininger, E. *Mod. Plast.*, 127 (June 1970).

FLUORESCENT DYES AND PIGMENTS

Fluorescent pigments, also known as daylight fluorescent colors, have been established for some time and are used in a wide range of applications. A daylight fluorescent pigment is a colorant which not only possesses the color it reflects but which also emits fluorescent light in the same color. The color observed from a fluorescent surface is a combination of reflected and emitted light, thus the colors appear extraordinarily bright.

The term *pigment* used in this context is actually imprecise, since by definition a pigment is an unextended inorganic or organic colorant insoluble in solvents of all kinds. With few exceptions, daylight fluorescent colors are not true pigments; they are solid solutions of fluorescent dyes in transparent synthetic resins which are finely ground to a particle size in the range of 2 to 5 μ.

These pigments could alternatively be described as transparent plastic materials colored with fluorescent dyes. Carrier materials are usually melamine formaldehyde or triazine formaldehyde resins, modified with aromatic sulphonamide resins. These materials are glass-like in character and are so brittle that they can be ground to an extremely fine consistency.

By varying the functionality of the resin constitutents either thermoplastic or thermoset resins can be produced; most pigment manufacturers supply pigments made from both types. The thermoplastic pigment melts more easily and is more heat sensitive but is more easily attacked by stronger solvents. It is, therefore, well suited for plastics like PVC, particularly plastisols and polyethylene, into both of which a thermoplastic pigment would melt very easily. The thermoplastic type has the edge on lightfastness and, in those plastics for which it is suitable, will melt during processing, obviating the use of powerful dispersion equipment. The thermoset pigment would be preferable with materials such as polypropylene or extruded PVC because of their relatively high processing temperatures – often above 200°C.

The dyes used in fluorescent pigments are mainly of the type 4-aminonaphthalimide (yellow) or rhodamine (red-blue/red) singly or blended together. These structures share both an internal atomic arrangement (called conjugated double bonds) and the capability of internal molecular changes (called resonance). These changes imply an increased ability of electrons to carry extra energy freely about in the molecule, which is an important characteristic of daylight fluorescent materials. There are no genuine blue or green fluorescent dyes; for these, phthalocyanine pigments combined with optical brighteners or with yellow fluorescent dyes are employed. The resultant shades are clearly more brilliant than those obtained with normal pigments; however, they never attain the luminosity of other fluorescent pigments.

Optimum fluorescent – apart from the fact that the fluorescent dye must be in molecular solution in the carrier resin – is dependent on specific dye content which, in practice, may be rather low. The surprising fact that fluorescence is not noticeably stronger with higher dye content – on the contrary, it is weaker – can be explained by the phenomenon of fluorescence quenching.

With increasing concentration of the fluorescent substance, an increase in fluorescence continues only until an optimum is reached. At that point the total light available for exciting the fluorescence is absorbed. If the concentration of the fluorescent substance is increased beyond the optimum, the fluorescence markedly declines, since at this point there is an increasing internal absorption of the fluorescent light by the molecules themselves. For this reason fluorescent dyes do not fluoresce when in undissolved powder form, i.e., in 100% concentration.

The maximum dye content of fluorescent pigments is set around 5%, to be at a optimum level. Fluorescence quenching occurs at high-pigmentation levels (e.g., 20 to 30%) in the case of thin PVC plastisol coatings and films. Combinations with other nonfluorescent colored or white pigments, and especially contamination with carbon black, also lead to a weakening or even extinction of the fluorescence, depending on the amount and inherent absorption. This is due to absorption by the nonfluorescing pigment or the contaminant, either in the wavelength region required for fluorescence excitation or in the emission region of the fluorescent dye. As a result, either the fluorescence is no longer fully excited or the emitted fluorescent light is partly absorbed. Combinations with nonfluorescing pigments therefore yield brilliant shades only if these pigments absorb as little as possible in the excitation region and in the emission region of the fluorescent dye.

Until recently one of the main disadvantages of fluorescent colors was their lack of heat stability, since most of these colors would degrade at temperatures above 190°C. Because of this, fluorescent pigments were limited to use in lower-melt plastics unless the scrap material in the molding operation was to be discarded. Considerably improved pigments are now available for incorporation into plastics such as high-density polyethylene, polypropylene, and polystyrene. These types of pigment can be processed at up to 250°C. Recently a new line of fluorescent colors, said to have 50°C more heat resistance, have been developed, eliminating or greatly reducing the problem of color degradation. It has also been reported that they have no formaldehyde odor, and that their resistance to sunlight has been improved.

In contrast to normal pigments, the lightfastness of fluorescent pigments is only moderate. This is a result of the poor lightfastness of the fluorescent colorants they contain; the carrier resins themselves are very stable to light. Fading in daylight can be slowed down somewhat by careful combination with organic, and possibly also inorganic, pigments of the same shade, although a reduction in the fluorescence has to be accepted. Durability is considerably improved in indoor use.

Although the heat resistance of fluorescent pigments has been improved sufficiently to make their use in most common plastics possible, the progress in lightfastness has not been so dramatic. However, the fluorescent colors generally last for the normal life of such items as toys. The industry has found that even in such areas as safety cones and flags, the color life is adequate for intermittent outdoor exposure. The only real area of concern has been in permanent signs which are continuously exposed to sunlight, and also coated fabrics such as those used for beach canopies. In spite of their poor lightfastness, the potential use of fluorescent pigments in plastics is almost unlimited, since fluorescent colors are up to four times as bright as the nearest corresponding nonfluorescent ones. That is why they are still used in signaling and safety signs, toys, coated fabrics for beach wear, high-visibility items, and fashion goods.

Fluorescent pigments are used primarily for transparent colors. In principle, the colors can be opacified by incorporating titanium dioxide or zinc oxide in the formulation; however, increasing the amount of the opacifier causes the colors to go pastel and begin to lose fluorescent brightness. Thus, as little as 0.05% of TiO_2, based on the weight of the plastic, causes a loss in fluorescent color, and 0.25% gives a very different pastel shade. An alternate approach to opacity is to use opaque, nonfluorescent toners of substantially the same hue as the fluorescent colors. The maximum amount of toner which can be used without appreciable loss in fluorescent brightness will depend upon the type and tones employed.

Because of the transparent nature of the fluorescent pigments, there is a potential problem with contamination. Contaminations, specially with black and brown color impurities, can quickly destroy the color and fluorescent effect of these pigments. It is important, therefore, that the equipment employed in the formulation and fabrication of fluorescent plastics be thoroughly purged or cleaned.

Some of the fluorescent dyes are sensitive to the action of plasticizers. Thus, when these pigments are used in plasticized plastics such as vinyl plastisols and films, the plasticizer must be chosen with care to minimize the tendency of the dye to bleed from one surface to another. Special pigments are available for use in this field but they merely minimize the bleed and do not prevent it.

One of the major factors inhibiting the widespread use of fluorescent dyes and pigments in plastics is the cost of coloration in thick sections. If exceedingly thick walls are involved, a film of fluorescent plastic can be laminated to a film of white or light-colored plastics. This is the method employed in the manufacture of most fluorescent traffic cones, where wall thicknesses may be as great as one-quarter inch or more. This system obviates using fluorescent pigment throughout the entire plastic article and permits reduction of raw material costs.

METALLIC FLAKE PIGMENTS

The use of metallic flake pigments in plastics is relatively new. Metallic flake pigments commonly recommmended for use in plastics are either aluminum or copper and its alloys. They exist in the form of shiny leaflets or thin flakes, and their value lies in the almost total reflection of light from their flat surfaces. This imparts a metallic appearance to the object.

Aluminum surfaces reflect very strongly throughout the visible spectrum, including the blue region. Aluminum pigments, as a result, produce brilliant blue-white specular highlights. At the same time, however, an element of grayness is present, particularly when aluminum of a very fine particle size is used.

The copper-based pigments, called gold bronzes, are actually brasses, i.e., alloys with zinc and minor amounts of aluminum ranging from 70% to nearly 100% copper. With increasing zinc content, the color changes from the characteristic red of copper to a progressively more yellow hue; oxide film on the pigment surface produces a mellower or deeper tone. The smaller-size grades, while less brilliant, do not give an impression of grayness, probably because there is no dominant bluish color element as there is with aluminum. Thus, despite lower total reflectivity, the copper alloys often produce a brighter metallic color than aluminum pigments of similar particle size.

The choice between aluminum and copper alloy pigment in the past often hinged on resistance to the conditions of processing and use rather than on the color effect desired. Aluminum pigments are not affected adversely by application conditions in a fluid media (plastisols and casting syrups); plastic milling, calendering or other high-shear processing quickly leads to dulling and graying, due to fracturing of flakes. In service they withstand most deleterious agents except strong acidic and alkaline solutions. Recently developed aluminum pigments are quite resistant to the shear encountered during milling, calendering and extrusion, and have improved resistance to acid and alkali.

Copper alloys were, until a few years ago, unusable in plastic applications because they tarnished rapidly under the influence of heat, light, moisture, and other conditions, as well as by chemical interaction with many plastics. Copper alloy pigments, developed in recent years, are apt to resist conditions of application and service in plastic applications, including shear, high processing temperatures, chemical interaction with plastics and prolonged exposure to light, moisture, and hydrogen sulfide. These improvements enable the choice of aluminum vs copper alloy pigments, even for durable applications, to be made on the basis of the most desirable aesthetic effect, without having to compromise aesthetics with durability requirements.

An essential requirement for metallic reflectance is the preferred orientation of the flakes parallel to the surface of the object. Hence, the orienting action of the production and finishing methods strongly affect the overall appearance. The color effect may be produced entirely by internal pigmentation of the plastic, or by a coating, or by a combination of both.

During calendering of plastic sheeting, plastic flow in the nips of the calender and stretching of the web during the cooling cycle contribute to flake orientation. In the extrusion of flat web materials and profiles, constricting plastic flow fulfills the same function. In both processes, avoiding nonuniform flow is as important as achieving desired orientation, because in the presence of metallics many flow marks show up glaringly which would go undetected in plastic of a solid color. This consideration becomes even more important, and often more difficult, in injection molding. With improper mold design the flow patterns may cause dark blemishes, due to flakes oriented edgewise to the surface, therefore rendered virtually nonreflecting. Blow molding and vacuum forming of plastics generally produce very good orientation due to the stretching which occurs in the last step, providing that misorientation or flow lines are absent in the parison or sheet stock, respectively. Compression molding generally produces enough surface flow to orient reflective pigments, but uneven prefilling of the mold cavity may cause surface material to flow sideways, away from a high spot; this reorients the pigment at different angles on both sides of the high spot, thereby producing a mottled effect.

Vinyl plastisols, used primarily in the production of coated upholstery fabrics, involve no significant shear breakdown; the heating cycle is generally short, but the temperature may be very high. Orientation as applied is generally modest; therefore, a solvent-based top coating or a finishing process of embossing or planishing should be used whenever very high brilliance is desired.

Post-molding of plastic surfaces is especially useful in conjunction with metallic colors. Orientation and brilliance are greatly enhanced during the production of a smooth planished surface or an embossed surface pattern. In addition, striking designs of subsurface pigment orientation, producing the illusion of a deeply convoluted metallic surface embedded in clear plastic, result from first embossing, then flattening the surface. In this case, the reorientation due to lateral flow-out of embossed peaks preserves the original embossing pattern, combining a striking three-dimensional effect with the easy cleanability and pleasant feel of a smooth surface. These methods have been used primarily on flat web materials, e.g., upholstery fabrics and rigid panel sheeting; they also present interesting possibilities in the production of extruded profiles for welting and trim, and the molding of handles and other small accessories. Flake pigments have also been incorporated onto plastics for decorative purposes. The flake can be sprinkled on the plastic, which may be coated onto paper and heated to embed the flakes, thus producing metallicized packaging film.

The visual sensation of metallic color involves separate perception of specularly reflected and nonspecularly scattered light from the same object. The color combinations useful to the stylist and plastics technologist may be classified according to objective interactions between metallic and colloidal pigment, due to the behavior of light and the subjective interactions connected with the visual recognition mechanism.

Aluminum pigments, reflecting a large portion of incident light at all wavelengths, are readily modified with transparent color of any hue. The difference between specularly reflected highlights and background color or nonspecularly scattered light is mainly one of intensity rather than hue; the highlights may be more silvery (i.e., paler or of lower chroma) than the background. A decrease in the degree of orientation, or addition of titanium dioxide or other light-scattering pigment, causes the nonspecular color or "flop" to become lighter while the highlights become less brilliant. Due to the strong reflectivity of aluminum in the blue region, reds tend to change toward magenta shades and yellows toward chartreuse. Very light pastels are grayed out by the combination of aluminum and opacifier. Strongly metallic colors of high chroma are produced by a metallic layer top-coated or over-laminated with transparent color, rather than by a single coating in which the aluminum would reduce chroma excessively.

An iridescent metallic color effect, which has no counterpart in the aluminum pigment system, is achieved by combining a copper alloy pigment with a color pigment of nearly complementary hue, causing the specular highlights to differ from the background color in hue as well as brightness. In other words, a combination of pale gold with blue shows a strong blue non-specular background color, shifting to bright yellow highlights at the angle of specular observation. In order to obtain this iridescent effect in thick plastic sections, a small amount of opacifying pigment, such as titanium dioxide, is added to scatter some light in a long average path through the colored medium, producing a strong effect on the background color; at the same time, enough light is specularly reflected to the eye to retain satisfactory brilliance of highlights. Under thin sections, a white backing layer fulfills the same function.

Virtually all of the metallic effects described here, whether based on aluminum or copper alloys, may be toned down in order to fit harmoniously into an overall color design. Addition of black or other low-chroma pigment produces very pleasing shades which, though subdued, retain the liveliness of metallic luster. For light color schemes, the additon of opaque, light-scattering pigments produces semipastels in which the metallic highlights have become much more subtle.

Metallic flake pigments have some shortcomings too. The effect of heat on bronzes has already been mentioned. Copper pigments may act as catalyst for light and heat degradation of some plastics, particularly for polyolefins. Excessively high-metallic pigmentation interferes with dielectric embossing and heat sealing, important operations in the construction of upholstery and soft trim.

Metallic pigments may become vulnerable to weathering agents, not only directly by the sensitivity of the metallic pigment, but also indirectly through the optical requirement of clarity imposed on the binder or plastic. Thus, even the new metallic pigments, which are very resistant to deleterious factors, must be used in a properly formulated matrix. A plastic impermeable to moisture will not only provide additional protection for the pigment, but will itself resist the blushing or whitening which occurs in some plastics upon absorption of small amounts of water. A degree of blushing that would pass unnoticed in a heavily pigmented solid color is readily visible and unacceptable with metallics, because it dulls the specular reflection. The same is true of darkening of the plastic due to atmospheric hydrogen sulfide. Both these effects, providing that they originate from a weakness of the plastic, are more critical for internally pigmented than for metallic-coated plastic.

In the small amounts used for toning metallics, some organic pigments which are generally rated as lightfast are quickly destroyed. This fading is more severe in combination with aluminum, which strongly reflects short wavelength radiation, than with copper-based metallics. As a result, the topmost zone of plastic receives a concentrated exposure of both direct and reflected light.

NACREOUS PIGMENTS

Nacreous pigments are pigments which impart a pearl luster. Their unique optical properties are derived from the plate-like shape and high refractivity of the pigment particles. The pigment platelets are transparent, generally colorless, and of high-refractive index. Interference pigments are special nacreous pigments in which color is produced by light interference (rather than light absorption), because the thickness of their platelet particles is in the interference color range. These pigments display both pearl luster and iridescent color. Nacreous luster arises in simultaneous reflection from many parallel layers. The effect is one of softness and depth in contrast to metallic luster which originates at a sharp plane of reflection.

The earliest known nacreous pigment is natural pearl essence, a suspension of the glistening, silvery crystals obtained from the scales or skin of fish. The synthetic nacreous pigments are inorganic products, but they share with natural pearl essence crystals the characteristics of plate-like shape and high-refractive index. The major synthetic pigments currently in use are bismuth oxychloride, lead hydrogen arsenate, basic lead carbonate, and titanium dioxide-coated mica. The last two compounds also serve as interference pigments.

For a discussion of the origin of pearl luster and iridescent colors the reader is referred to the two excellent articles given in the bibliography. In brief, the pearl consists of alternating layers of calcium carbonate and a proteinaceous material. Incident light is partially reflected from the top surface and partially transmitted. A second reflection occurs at the next interface, and so on, with the result that the reflection arises in depth to give the optical characteristics of the pearl. The pearl usually contains uniform layers of small thickness, which when suitably spaced give rise to color by the interference of light. The pearl and mother-of-pearl, therefore, display both pearly luster and iridescent color: the specific color depends on angle of observation as well as the thickness of the layers. It should he noted that the iridescent color is completely different in origin from that produced by colorants. In iridescent color no selective absorption of light waves takes place, as is the case with common pigments and dyes.

The synthetic nacreous pigments resemble pearl in that they consist of thin platelets, although not necessarily of stacks of alternating platelets. The sandwich or layered structure, which is essential for the realization of pearl luster, is produced when the pigment platelets are embedded in plastics of refractive index different from that of the pigment. In practice, the refractive index of common plastics and film-forming polymers is not far from 1.50, while the refractive indices of nacreous pigments are in the range of 1.60 to 2.50. Thus, a layered structure is produced, consisting of two polymer/pigment interphase reflecting surfaces which give rise to partial repeat reflections like a pearl's.

The reflections, arising from two reflecting surfaces, interact with another; the result of the interaction depends on the phase difference, itself a function of platelet thickness, refractive index, and angle of incidence. If the optical thickness of the platelet (geometric thickness $\times$ refractive index) is in the range of 200 to 600 nm, selective reflection and transmission can take place and the pigmented mass may appear colored. If the geometric thickness of the platelets changes, the color of the pigmented mass may also change. A characteristic of interference color is its dependence on angle of incidence. The color shifts with direction of illumination and observation, creating the play of colors which is described as iridescence.

The twin colors which are inherent in any sample of interference platelets make possible unique color effects. For example, a coating containing the platelets will dispay only the reflection color if the coating is over a black background; any transmitted light is absorbed by the background. The same coating applied over a white background changes in color, depending on whether it is examined at the specular angle (i.e., where the angle of reflection equals the angle of incidence) or at some other angle. The reflection color is seen at the specular angle; at all other angles the light passes through the platelets to the white background, is reflected back through the platelets a second time, and reaches the observer as the transmission color. A coating on a curved surface with a white background displays both colors at once, the reflection color being seen in the highlight and the transmission color elsewhere.

Interference pigments provide a variety of color effects through different combinations of interference pigment and background color. For example, yellow over black produces an almost metallic gold; over white, a light blue (the transmission color) with light gold highlights; over blue, a darker blue with gold highlights; over red, red with gold highlights (the transmission color being masked); and over yellow, an intense, lively yellow in which the color of conventional yellow pigment (absorption of blue and red light) is enhanced by reflection of yellow by the nonabsorbing interference pigment.

The intensity of interference colors is increased by a dark background. A white background reflects light which dilutes the particular reflection color of the interference pigment. Only a light background, however, permits both reflection and transmission colors to be seen simultaneously. These different characteristics are seen in nature in mollusk shells: the oyster shell generally has a white background and displays subtle colors; the abalone shell has a dark background and displays intense colors.

The nacreous pigment with the described platelet characteristics imparts high sheen and luster and only moderate opacity. A film containing the pigment is therefore translucent, and its appearance is affected by what lies underneath. An opaque white undercoat creates the appearance of white pearl. A translucent substrate, such as an "alabaster" glass bead, enhances the impression of depth. A black undercoat creates a somewhat metallic effect.

Nacreous pigments differ widely in their physical and chemical properties. Their use is, therefore, very selective and strongly dependent on the method of fabrication of plastic and the intended use of the plastic object.

The use of bismuth oxychloride nacreous pigment is not very common in the plastics industry. It is primarily used in PVC molded articles and film for applications where opacity is a primary consideration. Because of its high-refraction index, bismuth oxychloride has a higher reflectivity and opacity than do other nacreous pigments. Its pearl brilliance is, however, lower because its surfaces are much less smooth. Bismuth oxychloride, unusual among nacreous pigments in being light-sensitive, requires the use of a UV absorber to prevent graying of the material in light.

Lead hydrogen arsenate and lead hydrogen phosphate are used for soft pearl luster in plastics applications requiring good chemical stability. They are particularly suitable for casein-formaldehyde and extruded cellulose acetate articles, in which the commonly used lead carbonate may undergo chemical degradation and in which natural pearl essence may react with formaldehyde.

Basic lead carbonate provides the highest pearl brilliance of any nacreous pigment and is the only interference pigment at present available in the full range of first and second colors. The relatively complete control which can be exerted over crystal dimensions accounts for its versatility. An inherent disadvantage is its reactivity with acids and sulfide.

In spite of its reactivity with acids, basic lead carbonate is effective in applications where an appreciable concentration of free acid is unavoidable, such as casting in unsaturated polyester resins. The nature of the acid is as important as concentration. Special stabilized basic lead carbonate formulations have been devised for those situations in which excessive concentrations of acid or sulfide are encountered.

The heat stability of basic lead carbonate is high enough to permit its widespread use in thermoplastics, such as poly (vinyl chloride) film and both high- and low-density blow-molded polyethylene bottles. Heat stability is only borderline with other thermoplastics such as poly (methyl methacrylate) and polypropylene.

A relatively recent development is a new class of nacreous pigment in which each platelet contains three layers, the outer layers having a higher refractive index than the inner. Titanium dioxide-coated muscovite mica is in this category. The mica platelets provide the desired shape for a nacreous pigment, and the titanium dioxide layers the necessary high-refractive index. The three-layer structure, or sandwich, acts optically as if it consisted of two titanium dioxide platelets supported by an intervening layer of low-refractive index which makes no significant optical contribution. The product is non-toxic and extremely stable. The titanium dioxide is inherently more reflective than any other nacreous pigment platelet because of its high-refractive index, and opacity is high in spite of the presence of the noncontributing mica. Other advantages of this nacreous pigment are its excellent chemical, heat, and sulfide resistance. Furthermore, it is relatively resistant to fragmentation in handling; the mica substrate helps to withstand the shear forces which are encountered in plastic processing equipment such as Banbury mills and extruders.

The major current uses for each nacreous pigment are given in Table 12. Basic lead carbonate is used for a low-temperature application

Table 12. Typical uses of nacreous pigments in plastics.

Pigment	Uses	
	Plastics	Products
Basic lead carbonate	Polyester and Poly (methyl methacrylate)	Sheets and rods for buttons, serving trays, umbrella handles, decorative packing
	PVC film	Curtains, rainwear
	Injection-molded polystyrene	Jewelry boxes, combs, brushes
	Cellulose acetate	Eyeglass frames
	Blow-molded polyethylene	Containers for cosmetics
Natural pearl essence	Cellulose acetate	Eyeglass frames
Bismuth oxychloride	PVC	Boots, saddles, rainwear
Lead hydrogen arsenate	PVC	Shower curtains, rainwear
	Extruded casein	Buttons
Lead hydrogen phosphate	Extruded cellulose acetate and casein	Buttons
Titanium dioxide-coated mica	PVC film	Curtains, rainwear
	Injection-molded polystyrene	Containers for cosmetics

like cast film and for sheets made at moderate temperature. Extruded cellulose acetate requires higher temperature, therefore more heat-resistant pigments are often used. Natural pearl essence is the pigment for brilliance; lead hydrogen phosphate is for moderate luster.

SPECIAL EFFECTS FROM INCOMPATIBLE POLYMERS

Combining two polymers that are incompatible and have different refractive indices may produce nacreous luster. The chief example is polystyrene/poly (methyl methacrylate) mixture. The former has a refractive index of 1.59, the later one of 1.50. The molding of incompatible polymers produces closely spaced alternating layers which give rise to the nacreous appearance. When the major component is polystyrene, its mechanical properties are to some extent impaired; shock resistance, for example, is reduced.

Other combinations of incompatible polymers which display a nacreous effect include polystyrene/poly (vinyl acetate), cellulose esters with various thermoplastic resins and mixtures of incompatible saturated polyester thermoplastics. A related effect is reported for polystyrene with metallic salts of fatty acids.

In addition to the nacreous effect produced by a combination of two lightly colored polymers, such as a combination of off-white polycarbonate combined with a beige-colored plastic material. In this case both components can be the same plastic, or they may be different polymers. This area, however, falls under the art of fabrication rather than coloration and appearance modification.

BIBLIOGRAPHY

1. Felsher, H. C. and Hanan, W. J. "Brilliance of Plastics through Metallics." *Mod. Plast.* **41** (Sept. 1963).
2. Felsher, H. C. and Hanan, W. J. "Metallic Colors in Automotive Plastics." SPE Annual Technical Conference **XIII**. Detroit (1967).
3. Felsher, H. C. and Hanan, W. J. "Evaluation and Description of Metallic Colors." *SPE J.* **21**, 1317–1375 (1965).
4. Miller, H. A. "The Properties of Nacreous Pigments." SPE Annual Technical Conference **XIII**. Detroit (1967).
5. Alfrey, T. "Physical Optics of Iridescent Multilayered Plastic Films." SPE Annual Technical Conference **XIV**. New York (1968).
6. "Pigments" in *Encyclopedia of Polymer Science and Technology* **10**.

7. Komar, L. C. "Titanium Dioxide in Polymers." SPE Annual Technical Conference **XIV**. New York (1968).
8. Gray, T. J. "Daylight Fluorescent Pigments in Plastics." SPE Regional Technical Conference, Cleveland (1971).
9. Eckardt, C. and Hefti, H. "Comparison of Fluorescent Brighteners in the Production of Synthetic Polymer Fibers and Textile Industry." *J. Soc. Dyers Colour.* **87**, 365 (Nov. 1971).
10. Richter, J. "Fluorescent Dyes and Pigments." *Europlast.* **46** (Oct. 1973).
11. Lavell, H. H. "Applications for Fluorescent Pigments." *Pigm. Resin Technol.,* 20 (Apr. 1973).
12. Boloney, R. A. and Greenstein, L. M. "Optical Characteristics of Iridescence and Interference Pigments." *J. Paint Technol.* **44**, 39–50 (1972).
13. Londergan, M. C. and Spengeman, N. F. "Modern TiO_2 Pigments." *J. Plast. Technol.* **42**, 260–264 (1970).
14. Brooks, E. H. "An Investigation of the Opacity/Pigment L Relationships in Selected Thermoplastics Containing TiO_2 Pigment." SPE Annual Technical Conference (1970).
15. Wicher, A. "Plastics-Brighteners." *Plast. Eng.* **29** (9) 36–39 (1973).
16. Dyer, R. "TiO_2 Pigments in Plastics Interactions." SPE Annual Technical Conference (1971).
17. Manasso, J. A. "Titanium Dioxide for Polymer Applications." SPE Annual Technical Conference (1975).
18. Hickcox, R. T. "Compounding Fluorescent Rigid PVC for Improved Exterior Color Stability." SPE Annual Technical Conference (1976).
19. Marvuglio, P. "UV Screening Properties of Transparent Pigments for Plastic Beverage and Packaging Applications." SPE Annual Technical Conference (1976).

5

Selection of Colorants

INTRODUCTION

Chapter 4 suggested the large variety of pigments that are available to the plastic colorist. However, not all colorants are useful for all plastics and for all possible uses. The colorist must choose colorants suitable for the specific plastics he wants to color, the specific processing conditions the plastic requires for fabrication and the specific end uses of the plastic product. A selection based solely on hue, brightness, covering power, and tinctorial strength would in many cases lead to disastrous results. Thus, once the tinctorial properties of a colorant have been established, consideration must be given to its fastness properties, its resistance to heat, light, moisture, chemicals, etc., and its interaction with the polymer or with the stabilization system. These requirements are briefly reviewed in this chapter. Sometimes these requirements must be met before the colorant can even be considered. Sometimes, when it is known that a requirement cannot be met in full, minimum acceptable standards will be agreed upon together with an appropriate method for testing these properties.

COLOR CHARACTERISTICS

The color characteristics of a colorant include its opacity (or transparency), relative tinctorial strength, and optical effects (e.g., iridescence, metamerism, and dichroism) in addition to its spectrophotometric response. The importance of the optical properties of pigments is not evident to the novice, but it is the manipulation of these characteristics which often determines the success or failure of a color-matching attempt. These topics are discussed elsewhere in this book: here we are concerned with tinctorial strength and the covering power of a pigment.

The *tinctorial strength* of a pigment is defined as the degree to which the colorant imparts color to a standard white pigment. It corresponds to the color yield of a pigment. The manufacturers of pigments generally furnish information on the tinctorial strength of pigments in the form of their tinting strength ratings. These measurements are obtained by grinding a weighted amount of colorant into a lithographic varnish for a specific length of time. However, as the tinting strength of a pigment is largely dependent on the degree of its dispersion in the coloring medium, and as this procedure provides no tangible information on the dispersing properties of the pigment in plastics, it is not a reliable index of relative tinctorial strength for the coloration of plastics.

Although the degree of dispersion has a profound effect on the tinting strength and hue of the pigment, the primary basis of determining its tinctorial characteristics is the chemical and crystalline structure of the colorant. Some pigments are inherently weak in tinctorial strength: ultramarine blue, cadmium sulfoselenide, cadmium mercury pigments, iron oxides, molybdates, cobalt aluminate, barium strontium, and zinc chromates, etc., are examples. Some other colorants are inherently very intense in color, notably most of the organic pigments and dyes.

Closely related to tinctorial strength is the hiding power or transparency of a colorant. Dyes are used for coloring transparent plastics; inorganic pigments are normally used to color opaque plastics. However, there are some organic as well as some inorganic pigments available which can be dispersed so finely in the plastic that a transparent coloration is achieved.

HEAT STABILITY

Heat resistance of a colorant is defined as its resistance to change in color or performance as a result of heat encountered during processing

These changes may result from chemical decomposition, crystal change, or interaction of the colorant with other ingredients in the compounded plastic. Heat stability is a function of temperature and time, and may affect performance without affecting color.

Heat stability is becoming an increasingly important consideration in the choice of a colorant for plastics, because of the increasingly exacting demands of the industry. For instance, when the plastic industry was in its infancy, processing temperatures above 200°C were rare, but today temperatures of 300°C and above are common.

Plastics differ in their range of processing temperatures. Thus, casting and thermosetting resins require relatively low temperatures (although for a prolonged time); rigid and plasticized PVC and polyethylene copolymers need moderate temperatures – generally in the range of 170 to 200°C; low- and high-density polyethylenes, polystyrenes, and cellulosics are worked in the range of 250 to 280°C; polypropylene, polyamides, ABS, polycarbonate, etc., require a temperature above 280°C. There are some plastics which need exceptionally high temperatures for their processing: examples are fluorocarbons, silicones and newly developed, high-temperature plastics.

The time of compounding during extrusion of pigmented plastics may vary from a few minutes to 30 min., depending on the type of extruder used and through-put rate. Similarly, the injection cycle in molding processes may vary from product to product and machine to machine. Given a particular plastic, some fabricating processes will need higher temperatures than others. For example, the extrusion of polypropylene into fibers and film would require a melt temperature of about 300°C; while for injection molding or profile extrusion a temperature of about 250°C can easily be accommodated. Still further, the plastics of lower-viscosity grades can be processed at lower temperatures than the same plastics of high-viscosity grades. All these factors have to be taken into account in selecting a colorant for a particular plastic.

It is not entirely logical to separate the heat stability of a colorant completely from its compatibility in a particular plastic/stabilizers system. Heat undoubtedly contributes to the breakdown of an unsatisfactory colorant, but it is not always easy to trace the decomposition to the effects of heat alone. The environment is seldom chemically inert in a plastic; acid (casting phenols, aminoplasts, PVC, cellulosics), alkaline (molding phenolics, polyurethanes), oxidative (polyesters, polystyrene) or reductive (molding phenolics, aminoplasts, polyamides) conditions may exist or develop owing to the thermal breakdown of the system. The combined effect of temperature, time, and reactive medium frequently leads to marked color changes in the coloring matter. The degree to which a colorant withstands these conditions is termed its stability to processing or, more simply but less accurately, its heat stability. Because of environmental differences the heat stability of a colorant can vary from plastic to plastic.

Another heat-stability problem can occur in processes such as high-speed extrusion: they increase the temperature of the material excessively, not to a point at which the colorant is decomposed, but to one at which a partially soluble organic pigment is completely dissolved. For example, an increase in the temperature of PVC can cause a shift in color from red to orange or from orange to yellow. In addition, pigment will bleed and bloom from the plastic during storage and will have almost no resistance to fading.

In contrast to the generally poor heat resistance of organic pigments and dyes, inorganic pigments generally exhibit excellent resistance to heat. High-processing temperature, therefore, eliminates most organic pigments and dyes, and the palette of useful colorants for high-melting plastics comprises mostly inorganic pigments. Of the useful organic pigments stable at the processing temperatures of about 300°C, phthalocyanine green and blue and carbazole violet possess the greatest overall utility, followed by quinacridones, isoindolinones, and perinones. The greatest deficiency is noted in the red and yellow regions of organic pigments. This is certainly an area of a future innovative organic pigment technology.

Speaking generally, there are not more than ten pigment categories, organic or inorganic, which are satisfactory at working temperatures in excess of 400°C, and some of these would be questionable if the temperature were held in this vicinity for any great length of time. Consequently, the selection of pigments for plastics processed at elevated temperatures is a simpler task than for those worked at lower temperatures, because of the relative scarcity of pigments which will withstand the heat cycle.

COMPATIBILITY

For a colorant to be used successfully it must not react chemically with the plastic or with other ingredients of the plastic formulation. This inertness of the colorant-additive-plastic system is loosely referred to as its compatibility. The compatibility of a colorant with any particular plastic system is often associated with its heat stability, since a reaction may take place during high-temperature processing that would not otherwise occur.

A common problem of compatibility is the degradation of the colorant by the plastic, polymerization catalyst-residues, cross-linking agents, stabilizers, plasticizers, antistatic agents, and processing aids. Nylon and aminoplasts, for example, behave as strong acid-reducing agents when molten, discoloring certain organic dyes and pigments that are normally stable at the processing temperature. Similarly, phenolic plastics may act as mildly alkaline-reducing agents affecting the heat and weathering stability of certain other pigments. Alkali metals present in PVC emulsion polymers or certain stabilized grades of polypropylene will "base exchange" with the alkaline-earth metals of toners, changing their color from bluish red to orange. Metal-free phthalocyanines will complex with metals (notably lead) from other materials present in the formulation. The residual oxidizing catalysts in polystyrene – and, more so, the cross-linking agents in unsaturated polyester compositions – exert a strong deleterious effect on some organic pigments and dyes, and in some cases also on inorgainic pigments. Some inorganic pigments, such as ultramarines, are sensitive to hydrochloric acid. They will not be compatible with such polymers as PVC, chlorinated polyethylenes, and fire-retardant polyolefin compositions.

Some pigments and the impurities associated with them have damaging action on polymers. Thus, it is important to avoid incorporation of pigments containing copper and manganese (except copper phthalocyanines) in polypropylene, and zinc and iron compounds in the PVC compositions, sometimes even to the extent of avoiding the very useful red and yellow iron oxides. Cobalt and manganese pigments can produce odor and brittleness in polyethylene on aging, and cadmium yellow pigments have caused the formation of gelled particles in polyethylene; similarly, pigments cannot be used indiscriminately in polyacetals, because some pigments cause uncontrollable decomposition of the polymer during processing.

Another important aspect of compatibility of colorants in plastics is their interaction with heat and light stabilizers and other additives with their decomposition products. The polymer-pigment may be compatible as such, but the minor additives or impurities present in the plastic composition may react with the pigment. The reaction of sulfur-bearing pigments (cadmium sulfide, mercadmium pigments, and selenium yellows) with nickel-bearing stabilizers (such as those used in some film and fiber-grade polypropylenes), or lead and tin-bearing stabilizers (such as those present in PVC), are well-known. These reactions result in severe discoloration of the melt or, in minor cases, in a dull shade. Chrome pigments are normally contaminated with lead compounds: they cannot be used, therefore, in combination with sulfur-bearing stabilizers such as DLTDP and thiobisphenols used for the thermal stabilization of polyolefins. The same is true with neaple yellow, which contains lead as a part of its structure.

WEATHERABILITY

Weatherability refers to the ability of a pigment to withstand all forms of weathering, including visible and UV light, moisture, temperature, atmospheric oxidation, and chemicals in whatever combination they may exist during the life of the product. Weatherability traits of importance include colorfastness, chalking, and retention of physical properties. Weatherability is often a function of light stability, oxidation and moisture resistance of a pigment or pigment-plastic system.

The importance of weathering resistance varies appreciably with the product and its application, ranging from the one extreme (building panels, advertising signs, or car rear lights, in which pigments are expected to be stable to exposure for at least 10 yr. to the full sun) to the other extreme (insulation materials for cables which are sheated over and buried in ducts, to be exposed to atmosphere only about a dozen times in their lifetime).

Fading may be due either to a loss of color of the dyes or pigments, or to the plastic becoming white through surface breakdown of one sort or another. Inorganic pigments generally have excellent stability to light; organic pigments range in quality from poor to excellent according to their structure. Furthermore, a pigment of poor stability to weathering will appear to be affected more significantly in a formulation containing a white pigment, if only because in a transparent formulation the fading of pigment on the surface reveals colored pigment underneath, whereas in a formulation containing a white pigment the latter becomes apparent.

The lightfastness of colorants is commonly expressed in terms of a 1 to 8 scale; 1 being the poorest and 8 being the fastest. Most of the pigments offered commercially to the plastics industry are known to command high ratings of lightfastness. The quoted ratings for lightfastness are, however, not wholly reliable because the light stability of most of the common colorants in one plastic does not correspond with ratings quoted by the pigment manufacturer in another plastic. This is due to the strong interaction of many pigments with certain polymeric substrates, stabilizers, and other additives. A careful testing is necessary before a colorant can be recommended for production use.

Although some pigments show little or no effect on exposure to weather and light, others, though relatively good in resistance to light alone, fail rapidly when exposed to oudoor conditions. An example of this is cadmium yellow. When used in interior applications it exhibits excellent stability both in masstone and tint to both sunlight and artificial light. Outdoors, in tints or pastels, it fades very rapidly. Another example is lithol rubines: they are suitable for most indoor uses but not for any outdoor application.

Fading, darkening, or changes in hue are usually due to the presence on the pigment of reactive groups which are affected by moisture and chemicals in the atmosphere, specifically acids, alkalies, and smoke – the three major pollutants found in industrial areas. Cadmium yellow, being and inorganic sulfide pigment, fades when exposed to a combination of moisture and sunlight; lithol rubines, a class of monoazo pigments that are precipitated with metal salts, show poor resistance to both acids and alkalies.

The degree of fading is often dependent upon the concentration of the pigment employed. Pastel shades exhibit lower fastness ratings than the heavier shades. The degree of dispersion also has a profound effect on the light stability of a pigment.

Attempts have been made in the past to increase the lightfastness of colorants in plastics by incorporating UV absorbers in the formulation. The results have been disappointing, probably because the amount of absorber in the thin layer of plastic between the surface and the first pigment particles is too low to be effective.

Compromises may have to be made in using pigments as colorants in plastics. The pigment may have to be used in combination with other colorants, whites, extenders, or other components which could affect its resistance to light and weather. Light fading is often dependent not only on the concentration of the pigment employed, but also on the type of reducing pigments used at any particular concentration. The addition of white pigments, especially in the case of light-sensitive colorants, causes a reduction in lightfastness, the degree varying with different pigments. In practice, if a choice between two colorants with interior light stability has to be made, preference should be given to the one that darkens rather than fades on exposure.

RHEOLOGICAL PROPERTIES

Flow properties of polymer melts are sometimes profoundly affected by the presence of certain pigments in the formulation. This is due mainly to the stabilizer-pigment interaction or to catalytic decomposition of the polymer by the pigment under the influence of heat. The change of rheological properties through an internal lubrication action or plasticization is not known. Occasionally, the flow properties of a formulation are not grossly affected, but pigments may cause some secondary effects which can be traced back to some subtle changes in rheological properties of the melt. Thus, the poor drawability and the existence of thick and thin filaments in fiber production or random fluctuation of thickness gauge in polypropylene film production have been traced to the rheological effects caused by the pigments added to the polymer. These difficulties are probably caused by localized degradation of the polymer and the creation of micro-heterogeneities in the melt viscosity. Inorganic pigments based on cobalt and cadmium, and organic pigments based on phthalocyanine structure are known to have some of these effects on polyolefins.

MORPHOLOGICAL EFFECTS

Some pigments, by virtue of their nucleating power, may influence the morphological structure of the solidified plastic. These pigments influence the formation of finer crystallites in the polymer on cooling. Talc and graphite have this effect on nylon, causing it to become more rigid and less impact-resistant; phthalocyanine blues and greens have been known to cause the formation of small crystallites in polyethylene and polypropylene film, resulting in excessive shrinkage. Quinacridone reds are known to have morphological effects on polypropylene, considerably increasing rigidity and decreasing impact strength.

EFFECT ON MECHANICAL PROPERTIES

Pigments generally do not affect the mechanical properties of molded plastics of thick cross sections, but they can have a significant effect on the mechanical properties of articles with thin cross-section, such as film and fibers. In general, most inorganic pigments tend to impair the drawability and tensile properties of polyester, nylon and polypropylene fibers and film. This deteriorating effect of pigments may have several reasons: The pigment aggregates and agglomerates can serve as weak points in the polymer continuum; the physical properties can become impaired through the induced decomposition of the plastic in the presence of certain pigments; or the properties may be affected through morphological changes in the polymer.

FILTRATION CHARACTERISTICS

Excessive pressure rise in the melt filtration during pelletizing and extrusion of pigmented plastics is an ever-present technical problem in the processing of plastics, necessitating a frequent change of filter screen and causing a large amount of waste, high labor costs, and prolonged periods of equipment shutdowns. This problem is especially critical in the extrusion of fibers and film, because of the very fine filtration screens used. Filtration characteristics of a pigmented plastic formulation is a function of the degree of pigment dispersion and, probably more important, is dependent on the uniformity of dispersion. A pigmented melt with a few very large agglomerates may not perform in practice as well as one containing a large number of smaller agglomerates. All those factors which influence the degree and uniformity of pigment dispersion also affect the filtration characteristics of the formulation. Some of these factors are discussed in the next Chapter. As will be shown there, pigments differ in their dispersibility characteristics. Inorganic pigments generally disperse more easily than the organic pigments.

MIGRATION

There are two types of phenomena collectively referred to in the plastics industry as migration: *blooming* and *bleeding.* Blooming is defined as the migration of colorants from within a plastic to a free surface, bleeding as the migration of colorant from one plastic material, through an interface, into an adjacent plastic or solvent phase. As the mechanism of migration is the same in both cases, both phenomena can conveniently be considered as physically identical, al though the results are different: blooming showing itself in poor fastness to rubbing or loss of luster in clear plastics; bleeding results in poor fastness to solvents.

The worst instances of blooming and bleeding arise with organic pigments that are soluble to some degree in oils and plasticizers. Blooming is sometimes manifested as chalking or bronzing of plastic surfaces. The blooming phenomenon is of special concern in those cases where the product is to be subjected to heat, such as annealing, or is to be used in hot environments. Bleeding is clearly undesirable in a plastic composition if the product comes in contact with solvents or if the plastic portion of a product is used in contact with other plastic components.

The migration of colorants occurs only in those plastics in which the main restraining forces on the pigment molecule are not too great to allow the pigment molecule or particle to force its way through to the surface or interface. The rigidity and compactness of molecular structure have been cited as the major factors limiting pigment migration. Polar groups have some effect. Increasing quantities of plasticizer gives rise to a more open molecular structure in the plastic and decreases the interbonding of the polymer chains, increasing the rate of migration.

It has been noted that the pigment must be soluble in either the plastic or the plasticizer if migration is to occur. Since, in both blooming and bleeding, pigments diffuse from inside the plastic to the surface, the two processes are interrelated. The driving force in bleeding is simply the concentration gradient across the interface, but the driving force in blooming is less clear. The dependence of bloom on supersaturation (pigment dissolved at the molding temperature constituting a supersaturated solution at room temperature) has been thoroughly investigated by several researchers.

When bleeding or blooming occurs in a plastic, its severity is usually proportional to the concentration of the offending colorant. With plasticized PVC, however, this is not so. In this plastic, at high concentrations of pigment, these faults are not too serious, but as the concentration is reduced or the processing temperature raised they become serious. This effect has been explained in terms of preferential crystallization of the organic pigment on the surface of the plastic. At high concentration, the undissolved pigment particles serve as nuclei for recrystallization of the dissolved pigment on cooling. At lower concentrations of pigment however, all the pigment dissolves, a supersaturated solution is formed on cooling which crystallizes only at the surface of the plastic and bleeds readily on to materials in contact with it.

Summing up, the prerequisites for blooming or bleeding are:

1. A supersaturated solution of pigment in a polymer or a plasticizer-polymer system.
2. The ability of the pigment molecules to move in that polymer or plasticizer-polymer system.
3. The inability of pigment to crystallize sufficiently inside the polymer or plasticizer-polymer system.

On the first point, one can say that this will be achieved in all those pigment-polymer plasticizer systems in which the pigment, under the conditions of hot processing, is in true solution in either the polymer or the plasticizer or their mixture. Further, the degree of supersaturation (and hence the extent of bleeding or blooming under conditions of use) will depend directly on the difference between the extrusion temperature and the temperature at which the plastic is used, and on the temperature coefficient of solubility of the pigment in the plastic phase. The three conditions are encountered with many organic pigment-plastic systems.

In regard to the second point, the glass-transition temperature of the pigmented plastic system must be below the temperature at which the plastic is used if bleeding or blooming is to occur. This condition is fulfilled by many pigmented plastic systems, e.g., the polyolefins and plasticized poly- (vinyl chloride), even when the temperature at which the plastic is used is room temperature, and polyester and other systems are at higher usage temperatures.

The third point is the one that has previously been given the most attention and is fairly self-explanatory.

The new organic pigments with especially large molecules are resistant to bleeding and blooming because their large size has made them appreciably less soluble. Pigment migration from PVC can now be controlled, but it was not so long ago that bleeding in cables occurred to such an extent that the colorant migrated not only from the insulation through the cable sheath but also through the plaster on the wall. As the cost of the new high-quality pigments decreases and as the volume of manufacture increases, their use in plastics will become more widespread and so the problem of migration will tend to disappear.

PLATE-OUT

In the processing of PVC an unpleasant phenomenon called plate-out has been known for a long time. It is the occurrence of more or less strongly adhering deposits on machinery parts during the processing of PVC compounds. The deposits are found mainly in the sheeting production on mixing and calendering rolls. Plate-out is particularly unpleasant when colored compounds are processed, because the colored deposits are partly redeposited on the PVC and may simulate blooming. This is not real migration of a pigment, but the entrainment of certain pigment portions during migration, i.e., the exudation of certain components of the mixture during processing. Several stabilizers, lubricants and plasticizers are known to do this. That this process is not true blooming, is indicated by the fact that inorganic pigments, known to be completely fast to blooming, also exhibit plate-out. Blooming occurs not during processing, only afterwards; plate-out takes place on the processing machine.

It seems obvious that plate-out of a pigment in PVC is directly connected with its wettability. A pigment that has been completely wetted out by PVC should not exhibit any tendency towards migrating to the surface together with exuding components. If certain substances cause plate-out, they reduce the wettability of a pigment by PVC. This can occur in two ways: either they wet the pigment more effectively than does PVC, thus entraining it on exudation, or they cause a less thorough mixing (and consequently less wetting) of the pigment by reducing the shearing forces within the mixture. Which of these possibilities is predominant has not been clarified thus far.

In the case of chemically identical pigments, the pigment form with the larger specific surface (i.e., with the smaller mean particle diameter) is the one which has the harder grain and is less easily wettable. It should, therefore, produce a more pronounced plate-out. This is actually the case.

At equal pigment concentration, plate-out increases with rising processing temperatures. The simple explanation is probably that the component entraining the pigment migrates more rapidly at elevated temperatures. It may, however, also be interpreted that at a higher temperature, due to lower shearing forces, there is less thorough mixing and consquently reduced wetting of the pigment. On account of this it is entrained more easily. Both are probably superimposed.

At equal processing temperatures the plate-out of a pigment also depends on its concentration; it increases with rising pigment content. This is natural since, at higher concentrations, more pigment particles are entrained by migrating mixture components.

The conclusion to be drawn for practical application is that, for preventing or at least reducing plate-out, it is advisable to work at the lowest possible temperatures and to use organic pigments with a mean particle size as far removed as possible from the lower limit of grinding. Since, with increasing particle size reduces color strength as well as and transparency of the pigments are generally reduced, the middle of the road will always have to be taken.

COST

The cost of coloring plastics is often the foremost criterion for the choice of a colorant. The price per pound of the various colorants varies from a few cents a pound for some of the blacks to as much as $40 a pound for some of the specialized pigments with outstanding heat and light stability. The amount of pigment that has to go into the resin also influences the final cost. To achieve flesh color in vinyl, for example, as little as 0.01% colorant is required; to achieve opacity in a vinyl credit card, on the other hand, the loading can go up as much as 20% or more. The brightness or coloring power of the colorant is also a factor since, in some shades, less organic pigments than inorganics can produce a desired color.

Another important factor affecting the cost of coloring is the relative processing costs. Evidently, it is less expensive to start fabrication from mass-colored plastic pellets than it is to start from predispersed colorants and colorless plastic. The use of concentrate pellets falls somewhere in between these two extremes. Not enough is known of the financial advantages of using dry pigments in comparison with color concentrates for the coloration of plastics. The economy of any system probably depends on the scale of production, the range of colors produced, and the type of products.

There are certain intangibles which also must be taken into account, for example, inventory. Buying the colored resin directly from the supplier may leave one with an expensive stock of committed colors that cannot be changed. With dry coloring or masterbatch, on the other hand, inventory problems are simplified, since the user need only stock crystal resin and a few colors to meet his requirements. (It is interesting to note that several raw materials suppliers are meeting the cost problem in some new and unusual ways. For example, several suppliers offer plans based on pooling of orders to permit production on the scale that allows a reduction from a premium of 3¢ or 4¢ over uncolored polystyrene to 1½¢ or 2¢ a pound.)

Equipment costs are also involved. Many of the self-coloring techniques require added mixing equipment and additional time spent to achieve a desired dispersion. In cases where small-volume production runs are the order or frequent color changes on the same equipment are necessary, it may be less expensive to buy colored resin than to invest in equipment. Where large volumes and a relatively small number of colors are involved, the reverse is usually the case.

BIBLIOGRAPHY

1. Dills, W. L. and Reeve, T. B. "Effects of TiO_2 on Weathering of Polyolefins." SPE Regional Technical Conference, Cleveland (Jan. 1969).
2. Richter, J. "About the Plate-out of Organic Pigments in Rigid PVC." SPE Annual Technical Conference **XIV**. New York (1968).
3. *Encyclopedia of Polymer Science and Technology,* **10**, entry "Pigments," pp. 189–193. New York: *Interscience* (1968).
4. Bailey, E. J. G. "The Coloration of Plastics." *J. Soc. Dyers Colour.* (Dec. 1969).
5. Uzelmeier, C. W. "How Heat and Light Affect Pigmented Polypropylene." *SPE J.* **26**, 69–74 (1970).
6. Sonn, G. E. "Colorants for High-Temperature Polymers." SPE Annual Technical Conference, New York (1970).
7. Hill, G. R. "The Effect of Pigments on the Weatherability of HDPE." SPE Regional Technical Conference, Philadelphia (1972).
8. McDowell, W. "The Migration of Organic Pigments in Plastic Materials." *JSDC* (June 1972).
9. Volz, H. G. "Advanced Methods of Testing the Optical Properties of Pigments." *Prog. Org. Coat.* 2 1–34 (1973).
10. Turturro, A. et al. "Crystallization Kinetics and Morphology of HDPE Containing Organic and Inorganic Pigments." *Br. Polym. J.* 5, 129–139 (1973).
11. Holtzen, D. A. "Discoloration of Pigmented Polyolefins." SPE Annual Technical Conference (1976).
12. Tsunashima, M. "Practical Evaluation Methods of Cadmium Pigments for Application in plastics." SPE Annual Technical Conference (1978).

6
Dispersion of Pigments in Plastics

INTRODUCTION

To achieve even coloration of plastics, two criteria must be satisfied: the colorant must be evenly blended throughout the plastic medium and it must be adequately dispersed. *Dispersion* in this context refers to methods for wetting out and reducing in size the agglomerates and aggregates of dry pigments. *Blending* refers to the process of uniformly distributing pigments in the system after dispersion. The two terms are not synonymous, although they are often used interchangeably in the plastics industry. Occasions may arise when a pigment is well dispersed but unevenly distributed in the plastic, thereby showing speckles and flaw lines in the product.

Normally it is not difficult to achieve even distribution: it's in dispersion that problems often arise. It is to this problem that the colorist has to devote most of his attention in color formulation, leaving the question of proper blending to the responsibility of the production staff. In this Chapter we review the basics of pigment dispersion in plastics; the next Chapter is devoted to the actual incorporation of predispersed or easily dispersible pigments to plastics.

It has been stressed in the foregoing pages that the ability of pigments to impart color and opacity to plastic objects depends not only on their inherent ability to scatter or absorb light, but also on their particle size. Pigment manufacturing processes are generally designed to produce particle sizes that will yield optimum spectral characteristics. The so-called primary particle size may range from a few hundredths of a micron for most organic pigments to 1μ or more for some inorganics. However, for various reasons, these primary particles tend to stick to each other in the course of manufacturing, resulting in aggregates many times the desired particle size.

In the case of pigments precipitated from aqueous solution, the individual pigment crystals become cemented together in varying degrees due to surface forces, particle growth, residual salts, etc., during the filtering and drying operations, forming clumps of particles or aggregates. In calcined material some sintering or interparticle growth may occur, again producing aggregates. Although the pigment manufacturer grinds his products to fine powders, this powder may still contain some residual aggregates of varying sizes. In addition, the primary particles and aggregates may cluster together in loosely bound agglomerates which may be as much as several millimeters in diameter. Practically any pigment which has a particle size of less than about 0.2 μ will show appreciable agglomeration and will present some problems in dispersion.

The plastics industry is not troubled by either the primary crystallites or by the aggregates – both are too small to be seen by the unaided eye, and too small to introduce mechanical defects into moldings and sections. Only when they interact to produce agglomerates of a relatively large size do problems begin. If the user is to obtain full value from a pigment in terms of strength development, transparency, opacity, depth of color, brightness, etc., he must develop a pigment-binder system that breaks down these agglomerates and disperses the pigment to something approaching the particle size of the single crystals formed in the original precipitation or calcination process. This process of particle-size reduction and wetting out the pigment particles by the plastic or a suitable carrier is *dispersion.* Dispersibility describes the ease with which pigment aggregates and agglomerates can be reduced to an acceptable size in a continuous medium.

Pigments differ in how easily they can be dispersed. Pigments such as titanium dioxide and cadmiums are among the easiest to disperse; carbon blacks and iron oxides are the most difficult. Even within the same pigment family, there are marked differences in dispersibility because of the variation in manufacturing techniques. This often results in one pigment being superior to another under the given conditions of dispersion and fabrication.

Dispersion of a pigment in a plastic can be accomplished with varying degrees of effectiveness. Aside from characteristics of the pigment itself, two other major factors are involved, i.e., the type of plastic and the type of dispersion equipment available. A plastic's ability to wet-up pigments is dependent to some degree on the presence or absence of polar groups in the molecular structure of the given polymer. Hence, polyesters and similar partially polar polymers produce better dispersions than polyolefins which have no polar groups.

Plastic colorists and pigment technologists are aware that the degree of pigment dispersion in the given medium has a direct effect on the visual properties of the object. Transparency increases with reduction in aggregate size and approaches a maximum as the degree of dispersion nears the primary particle size of the pigment. Tinctorial strength, and along with it the economy of shade, is affected in the same fashion. What is, perhaps, not as fully appreciated is the fact that changes in hue also are directly related to the degree of pigment dispersion. Blue becomes clearer and greener as the dispersion improves, greens and reds become cleaner and yellower. Obviously, in any situation involving close color control from one lot to another, it is of paramount importance that the pigment dispersion be of consistent quality.

In addition to the problems of color intensity and hue, the degree of dispersion has a profound effect on the filtration characteristics of the polymer melt. A poorly dispersed pigment would lead to rapid clogging of the filter screen in extrusion and necessitate frequent interruptions in process. Additional direct effects of inadequate dispersion include high coloring cost, increased color control problems, poor mechanical properties of the extruded and molded articles (especially thin cross-sectional articles such as fibers and film), and a tendency to chalk and crock.

Practical aspects of pigment dispersion in plastics lie in recognizing the fact that the dispersion process is often costly and that reducing the pigment aggregates to their absolute primary particle size is not always required. The optimum particle size desirable for each pigment will vary and pigments susceptible to overgrind are known. In most cases, the optimum particle size of pigment dispersion for good optical properties lies in the range of 0.3 to 0.8 μ. Even this range of particle size is often difficult to achieve in plastics without incurring excessive costs. The criteria of optimum dispersion, which is important in theoretical discussions of the subject, are in practice equivalent to those of acceptable dispersion. Three important questions in this context are: 1) What is an acceptable dispersion? 2) How can it best be achieved? 3) How can the quality of dispersion be evaluated and controlled? A substantial part of the color technologist's time is spent answering these questions for every formulation he evolves. The answers are often subjective and judgmental.

Acceptable dispersion is determined by the relative importance of appearance and performance traits that can be affected by pigment particle size and particle size distribution. These traits are surface texture, specks, gloss, hue and strength, opacity or transparency, processability, cost, physical strength, electric resistivity, and weathering. Performance and processing are affected by pigment particle size in a different manner. Thus, particles 30 μ or larger are primarily responsible for surface imperfections, speckling and such processing problems as plugging of filter packs during extrusion. Particles in the 15- to 30-μ range seem to contribute most to gloss problems, whereas particles as small as 5 μ influence physical and electrical properties in critical applications. Still smaller particles are generally needed to realize the optimum color strength, opacity or transparency: it is in this area that the question of dispersion cost vs added value most frequently arises, and it is in this area where quality control is often a problem.

THEORY OF DISPERSION

Dispersion is usually achieved through a combination of three mechanisms:

1.	Initial wetting	formation of pigment/carrier mixture
2.	Size reduction	breaking up of aggregates and agglomerates
3.	Intimate wetting	displacement of air from pigment surface

Initial wetting is essential to every dispersion regardless of quality. As a minimum, it requires that pigment and carrier be sufficiently well mixed and have sufficient affinity for each so they will not separate when further work is applied to the system. The mutual affinity, compatibility, or wettability of two materials can be increased through a change in the surface characteristics of either or both, by use of surfactants. Pigments, being of many different chemical types with different surface characteristics, will vary in the rate at which they will wet-out in a given system. Sometimes this is the controlling factor in the overall rate and quality of dispersion that can be obtained regardless of the processing that follows. This varying nature of pigments also explains why no single surfactant is ideal for all dispersions. The importance of initial wetting is often underestimated because of the simple means by which it is usually obtained. Yet, initial wetting is not only essential, it often controls the quality of the final dispersion.

Studies of the processes which mix pigments and plastics together have generally excluded the size-reduction mechanism from consideration, but the size reduction invariably occurs along with wetting. Particle-size reduction requires that sufficient mechanical energy be brought to bear on the particles to overcome forces holding them together. This energy is usually in the form of shear stresses developed in the polymer, which ruptures the agglomerates. When these stresses (the magnitudes of which are determined by the viscosity of the polymer and the mixing conditions) are greater than some threshold value (which will depend upon the characteristics of the agglomerates and aggregates), size reduction will take place. If the stresses are smaller than required to overcome the strength of the particle, it will not be dispersed. Because the consistency and *tack* of the polymer influence the effectiveness with which the mechanical energy can be transmitted to the particle, predispersion in a medium other than the base polymer is often considered.

The bond energies between primary particles in pigment aggregates and agglomerates vary considerably in strength. Variations in manufacture and treatment also produce different combinations of weak and strong fractions. For this reason, one pigment may yield better results than another at a low shearing stress but be inferior at higher shearing stress. Ratings at several levels of stress for each dispersion attribute of importance are needed for a comprehensive evaluation of ease of dispersion.

Intimate wetting is the process of replacing air at the pigment-air interface with a vehicle. This is of great importance in color-pigment applications requiring high transparency or maximum chroma in dark shades. Haze and reduced chroma produced by light scattering at plastic/pigment interfaces if the wetting is incomplete. Effectiveness of shear transmission in the dispersion process also depends to a great degree on the amount of intimate wetting obtained. Shear provided by dispersion equipment through a fluid medium would have no effect on a pigment aggregate unless a bond exists between the pigment surface and the medium, and this is accomplished only through intimate wetting.

The same considerations discussed under initial wetting also apply to intimate wetting, i.e., a need for compatibility and the role of surfactants. However, intimate wetting is made more difficult by the much smaller particle sizes involved, particularly for organic pigments with their high-surface area and microscopic interstitial pores. Such pigments often require extended dispersion cycles to achieve desired results.

While many attempts have been made to study pigment dispersion, most have been restricted to the fields of paint and printing ink, and few relate to the plastics media. Irrespective of the medium used, dispersion levels have been shown to improve with increasing mixing energies. These are normally achieved by increasing the power or the speed of mixing, but they may also be accomplished with longer mixing times.

Dispersion is an exponential disintegration process analogous to radioactive decay. The amount of dispersion taking place at any time (i.e., the rate of dispersion) depends upon the number of dispersible particles present, their strength, and the maximum of shear stress. If all particles are of equal strength, the rate of dispersion is independent of time, although the total change in dispersion is not.

The total change in dispersion, and hence the ultimate coloration developed, is dependent upon the energy of the mixing process. As the time of mixing is increased, the value for the total change approaches, asymptotically, a maximum level corresponding to that achieved by mixing for an infinitely long time at that level of shear.

But long processing times do not compensate for a lack of intensity of shear, and no matter how long the process is extended, low shear levels will not generate good color development. This relationship has been observed with many different pigments.

Dispersion becomes progressively more difficult as particle size decreases. At some threshold size, any further increase in dispersion ceases completely. As yet there is no means available by which the strength of these small pigment particles may be measured, and therefore no definite idea may be formed of the size of the smallest dispersible particle or of the way in which this changes with the ambient rate of shear.

Since dispersibility varies with agglomerate size, it is to be expected that each size would disperse at a different rate. The observed rate of dispersion in a population of particles containing a range of different sizes will thus be given by the average of all the individual rates of dispersion. Since the rate of dispersion for small particles will contribute very little to the overall observed rate during the early stages of mixing, the early rate of dispersion will depend on the number of large particles. As the mixing cycle is extended and the large particles disappear, the small particles will become more dominant, and the rate of dispersion will fall. This may or may not be noticeable, depending upon the level of shear and the difference in dispersibility between the large and small particles.

As might be expected, the level of dispersion achieved with any mixing process depends upon the energy required to disperse the agglomerates and the ability of the process to supply that energy. The two aspects of this relation may be examined separately.

The mixer used for a compounding operation will, by the nature of its design, determine the maximum level of shear which may be developed within any given polymer at any operating temperature. It is well-established, for example, in the plastics compounding industry, that every mill develops a specific energy which in turn limits the level of dispersion finally achieved.

Design will also control the residence time (that is, the time for which the agglomerates are exposed to shear) as well as the proportion of that time during which the agglomerates are exposed to high rates of shear. Time must be allowed, in any mixing process but especially in batch mixing, to ensure that all the agglomerates which are present pass through the zone of highest shear. Other variables limit the shear developed during mixing; these include the operating temperature, the nature of the polymer, and the mixer speed. Higher speeds raise the energy input as well as pass material through the zone of maximum intensity more frequently. Thus, more agglomerates are dispersed more quickly.

The theoretical discussion on dispersion mixing can be summarized as follows:

1. Dispersion is a combined effect of size reduction and wetting-out of pigment particles by the carrier.
2. High shear promotes dispersion.
3. There is some threshold stress below which no dispersion will occur.
4. Larger agglomerates will disperse more easily than the small ones in the early stage of dispersion.
5. The total change at any time in a dispersion process is dependent upon the time of mixing.
6. The ultimate level of dispersion achieved depends upon the maximum shear available in the mixer. The higher this rate of shear, the greater the change in dispersion.
7. The maximum level of dispersion is approached asymtotically as mixing proceeds. Evaluation of dispersibility of pigments in plastics shows that dispersion is a first-order process.
8. Pigments differ considerably in their response to shear: in general, inorganic pigments disperse more easily than organic pigments.

DISPERSION METHODS

Having discussed the principles of the dispersion process, at least in general terms, the question of how to obtain it can be considered. The basic dispersion processes for pigment-plastic systems are:

Dry blend – low- to high-intensity mixing of pigment with dry resin particles. Dispersion primarily obtained through impact and attrition.

Melt shear – high shear transmitted through fluxed resin via kneading action.

Liquid – impact and/or shear in liquid carriers (e.g., plasticizer low molecular weight polymers) or solutions of polymer or monomers.

In practice, a combination of these processes is often employed to take advantage of the merits of each. For example, intensive dry blending before melt dispersing can improve dispersion quality as well as reduce overall cost, and predispersion in plasticizer is often the least expensive means of obtaining high-quality dispersion of some colorants in vinyls. The development of dispersion procedures by any of these processes has been largely empirical, depending heavily on the past experience a formulator has had with pigments, dispersing equipment, and resin systems. However, application of the dispersion theory should suggest ways in which these procedures can be improved.

DRY BLEND DISPERSION

Dispersion by dry blending alone occurs largely by impact and attrition. Except when followed by other dispersion processes, little opportunity exists in later processing to achieve the intimate wetting needed for some critical applications. Nevertheless, techniques and equipment have been developed that permit production of good-quality dispersion by this method. The most important of these are nonfluxing high-intensity mixers and use of particulate resin in pigment-resin coblending.

The need for equipment that will impart high velocity or impact to pigment particles is obvious, since aggregates and agglomerates can be fragmented only by exceeding the energy level with which they are bound. One common method is ball-milling the pigment with the plastic powder at low temperatures. In general, dry-blend dispersion has been more effective at low-pigment loading than at high, primarily because of the caking tendencies of many pigments and extenders.

Pigments vary not only in the energy required to break up aggregates, but also in their susceptibility to overgrinding. Fracture of the primary particle size (as happens with lead chromates and cadmiums) or compaction into aggregates larger than were initially present (many organics) are the two most frequent results of overgrinding, both of which produce large and undesired changes in color and often in performance. Some pigments have also been known to change crystal phase, with an accompanying sharp color change, when dry blended too vigorously with certain resins. These changes are, of course, in addition to the commonly encountered problem of discoloration caused by abrasion of the equipment during grinding.

The value of using particulate resin in preference to larger granules or pellets seems to lie in the significantly improved initial wetting obtained. The importance of this should not be underestimated, even if used only in preparation for melt shear dispersion. Experience shows that pigments are more easily and completely incorporated into a resin when they are attached to the resin surface prior to melting. Finely divided resin not only provides more surface for coating, which permits working at higher pigment concentrations, but also puts the pigment in more complete, intimate contact with the total resin mass so that shear is transmitted immediately and more effectively in the vital initial stages of fluxing. In other words, no large localized regions of high-resin and high-pigment content exist alongside each other. If they did, inherent differences in consistency and melting temperature would undermine the quality of mixing obtained.

It may seem illogical that a resin size that will significantly reduce impact stress when hitting a pigment particle will still result in higher-quality dispersion. However, this attests to the very real value of initial wetting to aid both size reduction (by shear) and more complete wetting in later stages. Of course, the pigment must adhere to the resin well enough to survive intermediate handling. Additives that will render the resin surface more wettable are often used to improve this bond, and use of resins with a highly porous surface has been suggested. Methods of increasing electrostatic attraction between pigment and resin might also be considered, since this appears to be the basis for good adhesion in some systems.

MELT SHEAR DISPERSION

As its name implies, melt shear processes depend on transmission of high shear through resin to achieve dispersion. This is the most-used method of pigment dispersion in plastics. Melt-dispersion conditions are usually best at, or very near, the softening point of the system because of the high-internal shear that can be developed at this point. At higher temperatures the process becomes more one of mixing than dispersion because of the greater fluidity. Opportunities may be available in the more complex formulations to withhold one or more ingredientrs (lubricant, plasticizer, low-tack oil) to improve the working consistency of the compound for dispersion; they can be added at the completion of dispersion. Similarly, highly efficient additives or extenders that increase consistency through surface absorption have been helpful. In line with this approach, dispersion at high-pigment loadings, i.e., concentrate levels, is the very essence of the well-known and highly regarded dispersion process used to provide high-quality dispersions to the vinyl, lacquer and cellulose acetate fiber industries.

As mentioned earlier, the effectiveness of shear transmission to a pigment particle through an intermediate fluid film will depend on: the quality with which the fluid wets the pigment and equipment surfaces; the viscosity of the system under actual dispersion conditions; and the internal cohesive strength of the fluid (referred to in some industries as its tack). Each of these, then suggests directions that can be taken to improve dispersion. Tack is an inherent characteristic of the resin; the only latitude in dealing with it is to either modify the resin or select an alternate, compatible carrier as a dispersing medium if tack severely limits dispersion in the base polymer. Wetting is a function of both the fluid and the surface to be wetted. Again, surface-active agents can often be used to improve wetting or an alternate carrier may be available offering better wetting characteristics. Viscosity is more within the practical means of the compounder to control, but first he needs some basis for determining what level is desired, and also to recognize that viscosities are likely to fluctuate widely during processing of thermoplastics.

All melt dispersion equipment for plastics depends on the utilization of hydraulic shear to break down pigment aggregates. This shear is, of course, directly proportional to the viscosity of the polymer which in turn is directly dependent on the temperature involved. The temperature developed in milling a plastic is a limiting factor in many operations since it reduces the viscosity on which the shear depends.

Three general types of equipment are in use today: two-roll mills, Banbury mixers, and compounding or milling extruders.

Two-roll mills are excellent for producing good pigment dispersion. Through careful temperature regulation of the rolls it is possible to maintain the viscosity of the plastic mass indefinitely, within a desirable range for producing the shear necessary for dispersion.

The Banbury mixer is designed to produce intensive mixing at high viscosities. A relatively high-power input is concentrated on a small volume of material. The result is extremely high shear and rapid temperature build-up. The increase in temperature results in a pronounced drop in the viscosity of the mass with resultant loss of shear. This is the limiting factor in this equipment, as dispersion must be accomplished very quickly before the mass loses the viscosity necessary for shear. Since very high-shear rates exist in the early stages of this process, it is possible to do a reasonably good job in the time usually available.

The two-roll mill and the Banbury mixer are both capable of producing good dispersion. It is usually necessary to use the Banbury in conjunction with a two-roll mill; this combination probably produces the best pigment dispersions in plastics that can be obtained by mechanical means.

The compounding extruder is similar to a two-roll mill in principle, with the important exception that with the extruder the material is in the working area for a limited time. Here again, dispersion is dependent on the development of suitable shear rates. This can be accomplished through proper screw design and temperature control in the compounding zones. Various screw designs are available for compounding. In all cases, fairly close clearance between screw surface and extruder barrel is desirable to generate the shear area so necessary for good pigment dispersion. The screw design should contain a large proportion of metering area. Some areas of the screw are frequently left without flights to further increase the area of minimum clearance between the extruder barrel and the screw. It is possible to do a fairly good job with an extruder – particularly if the pigment and polymer have been subjected to intensive mixing beforehand.

LIQUID DISPERSION

Liquid-colorant addition to plastics during processing has become firmly established in many resin-fabricating operations, and has a quite excited projected rate of growth. Consumption of liquid colorants for 1976 has been estimated at 2500 metric tons with a projection for 1980 of over 10,000 tons. The advantages of liquid colorants relate directly to better economics and the ability to automate operations. Technological advancements relate to improved product quality. In addition, the concept of colorant addition in a liquid carrier has been expanded to include the simultaneous feeding of many other plastics additives such as blowing agents, antioxidants, UV stabilizers, and lubricants, all in a single operation.

In liquid colors, the dispersion of pigments is made in a liquid carrier, which may be a plasticizer, polymer solution, low-molecular weight polymer, or a combination of these. Carriers for liquid colorants are selected for compatibility with a wide range of thermoplastic resins and plastisols. In fact, liquid colors are now available whose carrier systems are compatible with virtually all widely used thermoplastics, including many engineering resins such as polycarbonates, phenylene-oxide-based resins, nylons, polysulfones, and others. This degree of compatibility results from several specific properties of the carrier system, whereby there is a small degree of chemical compatibility that allows the system to bond to the polymer chain. However, the major degree of compatibility is obtained by an emulsification process that produces a high degree of syntactic abilities in the resin matrix. The total effect is to provide a carrier system with a high degree of stability for carrying colorants into a resin without significant effect on the resin's physical properties.

The carrier system for liquid colorants involves other considerations relating to product performance. One characteristic involves the optimization of pigment dispersion in the finished liquid colorant. The technology here consists of mathematically computing the required surfactant blends for a given pigment to permit the maximum amount of solids content with the minimum of residual surfactant in the dispersion. This process is described by the HLB system (hydophile-lipophile balance) built on the thoery that numerical designation can be calculated to reflect the type of behavior expected from a surfactant system. Then, coupled with the empirically derived established requirements of known colorants, pigments can be dispersed for their optimum effect while maintaining other important performance characteristics of the overall liquid-colorant dispersion.

The approach to pigment dispersion in a universal carrier has led to the ability to create liquid colorant dispersions to accommodate virtually all colorants used by the plastics industry. Hence, a formulator can meet the highest criteria for coloring requirements, including those of the FDA and USDA.

A greater variety of dispersion processes are available for liquid systems because they are easier to handle and offer greater latitude in formulation. The mode of energy transfer (i.e., impact, attrition, or shear), still dictates the level of fluidity needed. Liquid pigment dispersion technology has borrowed a lot from paint technology. All those technological processes which are employed by the paint industry are important in the preparation of liquid dispersions. Shear processes are common, as they perform best at high viscosity.

At high loadings, particularly, pigments can have rather pronounced non-Newtonian effects on viscosity so that viscosity measurement on a millbase at rest may give little insight into its actual consistency during milling. Furthermore, viscosities of millbases containing high-surface area pigments tend to increase sharply during dispersion as the pigment surface is wetted, so that viscosity adjustment may be required during the grind to maintain proper working consistency.

The quality of initial wetting in liquid dispersion is easily measured by its mix-in time, i.e., the time required to form a coherent mass, or, in the case of a three-roll mill dispersion, the time needed to eliminate tailings and separation. Surfactants can be particularly effective at this stage of dispersion. Progress in particle-size reduction is generally followed by fineness or texture measurement on a grindometer scale; however, the compounder is again reminded that this measurement does not necessarily correlate with development of other dispersion attributes of importance, e.g., color, opacity, haze, etc., which involve particle sizes that cannot be measured by this method.

Principles of millbase formulation and equipment operation to achieve optimum results in liquid shear dispersion are much the same as already described for melt shear processes. Literature sources are available that describe liquid dispersion processes in greater detail. Reeves and Zabel (see Bibliography) summarize their efforts to determine optimum dispersion conditions for three representative pigments in plastisol and calendered vinyl systems as follows:

1. Maximum strength and minimum speckling via the plasticizer predispersion route are obtained when:
   (a) The pigment is predispersed under conditions most suitable for good fineness.
   (b) The resulting paste is compounded into vinyl at a consistency conducive to uniform distribution in the vinyl.
2. Best results in both plastisol and calendered vinyl were consistently obtained from sand and three-roll mill dispersions regardless of the pigment or plasticizer involved, provided the formulating rules described above were followed.
3. Cowles, Kady, and even change-can dispersion may produce satisfactory results in specific instances, but the performance is highly dependent on the pigment and subsequent conditions of compounding into vinyl (e.g., heavy duty milling may be necessary to eliminate speckling).
4. Pigments differ considerably in their response to conditions of predispersions, e.g.:
   (a) The phthalocyanine blue varies in strength in calendered vinyl; whereas, the quinacridone red strength remains virtually unchanged.
   (b) Speckling of the blue is relatively minor, except at high paste consistency, in both plastisol and calendered vinyl; whereas, the red and scarlet tend to speck badly unless properly predispersed.
5. The plasticizer type has no significant effect on strength or speckling when the paste is formulated to yield optimum results. Production rates will vary, however, as a result of plasticizer influence on paste consistency.
6. Surfactants are particularly useful as a means of reducing paste consistency, with resultant improvements in strength development and speck elimination. However, they may cause discoloration on later heat processing. Most of the best surfactants for plastisols dispersions are cationic or amphoteric.

Rheological factors are paramount in determining the pigment loading of a liquid dispersion. The limiting factor in this connection becomes the viscosity that allows for proper incorporation into plastics, with the mixing equipment at hand. For plastisols, the optimum viscosity is a viscosity similar to that of the plastisol. For incorporation in thermoplastics, high viscosities can be tolerated.

In any discussion of viscosity, one cannot overlook the effect of the surface treatment on the pigment. It is different for each pigment manufacturer and can result in dispersions of different viscosities. Resistance to flow of liquid dispersions may sometimes become a problem. In this case, a reduction of pigment concentration generally alleviates the problem. "Short" and "buttery" dispersions are extremely difficult to incorporate into a plastisol. This problem arises because of some physical interaction of pigment with the dispersion liquid – probably a swelling of pigment particles is involved. Benzidine pigments are notorious for this.

The plastics industry is gradually turning to the use of liquid dispersions. Liquid pigment dispersions were introduced some years ago, with disastrous results. One reason was that the wrong dispersion media were chosen; also, the early colorants caused plating out on the screw. Another reason the liquid-colorant concept developed such a bad reputation was that one of the major areas of interest, polyolefins, was completely neglected in the early days.

Liquid dispersions utilize conventional high-performance organic and inorganic pigments dispersed in a liquid vehicle system which is essentially nonvolatile at normal molding temperature up to 400°C. Like solid dispersions, the vehicle becomes an integral part of the resin matrix. The vehicle is capable of carrying high loadings, in the range of 60 to 70% pigments. Because of the nature of liquids, the dispersion characteristic of pigments in liquid systems is much more important than in solid-type systems.

Some concern about the possible migration and exudation of the liquid vehicle on the surface of the pigmented plastic part still exists among the plastics technologists. Carrier migration has been, in fact, a source of trouble. New liquid dispersions are reported to be free from this defect and the recent experience of the industry has confirmed these claims. The coloring costs based on using liquid dispersions are competitive with the costs of dry dispersions.

EVALUATION OF DISPERSIBILITY AND QUALITY OF DISPERSION

If pigments are to be used efficiently, more information about their behavior in a dispersing environment is required. Several attempts have already been made to devise methods of measuring the quality of dispersion and the way it changes with mixing. Most of this effort has centered on the "tinting strength" test, although a method based upon particle size analysis has recently been developed.

The tinting strength of phthalocyanine blue in a white paint base has been shown to be directly related to its particle size. This provides a technique which may be applied to a plastic media, in which a standard white base is mixed with the pigment being examined. The resulting depth of shade, or tinting strength, is proportional to the level of dispersion of the colored pigment. A comparison is usually made between the sample in question and one or more controls, which were made under conditions especially chosen to give high or low dispersion. For an exact comparison of shades, spectrographic curves should be used, but direct colorimetric measurements are usually sufficiently precise to be used on most occasions.

There are, however, a number of associated problems. The level of dispersion can have a pronounced effect upon the shade of the final color and upon the transparency of a colored pigment. Since the amount of white pigment is always used so that the total amount of scattering should remain almost constant. It is assumed that this pigment remains unaltered during subsequent processing. Finally, large colored-pigment particles appear as nearly black specks in the mixture and therefore make only a small contribution to the color.
color.

There are other methods by which the size characteristics of pigment particles may be examined, although most are rendered useless by the presence of the solid polymer matrix which surrounds the particles in plastics. Of course, a microscope can be used, as it has been in the past, despite the inherent tedium and subjectivity normally associated with it's use. The recent appearance of automatic particle counters has largely removed these objections and have assisted in the development of procedures which count and size particles at high speed. There are several particle counters currently available, but only the *Quantimet* has achieved wide acceptance.

The measurement of the degree of pigment dispersion is a difficult task. Only the large particles can be observed under an optical microscope. One convenient and accurate measuring method is to prepare a 5-mil pressout of the colored plastic. A small portion of this can then be mounted for viewing in a regular optical microscope at 400X. Use of a graduated scale superimposed on the field facilitates estimation of aggregate size. Direct observation of films or pressed-out plaques also serves as a qualitative guide. Grossly poor dispersions appear to be hazy and even to have visible color particles present.

A qualitative estimate of the degree of dispersion can also be obtained indirectly by measuring those properties of the colored plastics which are dependent on the presence or absence of large particles of pigments. Some of these properties are: flow of melt through extremely small apertures, electric faults in a wire coating, clogging time of a filter, and the physical strength and integrity of fibers and thin films.

It has been mentioned earlier that the appearance attributes of a pigment often develop at different rates during dispersion. Hence, measurement of one does not necessarily indicate the value of another. In all studies of dispersibility of pigments, the rate of shear and time of shearing must be kept in mind as important factors determining the rate and quality of dispersion.

PREDISPERSED PIGMENTS AND CONCENTRATES

A number of easily dispersible pigments, predispersed colors and color concentrates has recently become commercially available. A few remarks are, therefore, in order to describe these forms.

Predispersed pigments consist of pigment particles held in a stable open lattice formed by an oil-soluble material, such as low molecular-weight polymers. An important feature of predispersed colors is that they contain a high proportion of pigment by weight. During dispersion in the plastic, the lattice collapses and rapidly dissolves in the plastic, releasing a fine pigment dispersion at low pigment concentration in the plastic. The carrier may be liquid (liquid dispersions) or solid (solid predispersed colors). Furthermore, the liquid dispersions are available as low viscosity materials (for plastisols) or high viscosity pastes (for thermoplastics). Similarly, the solid predispersed colors are available as pellets (to be used in pelletized thermoplastic materials) or powders (to be used in powder resins). The carrier is generally a mixture of one or more polymeric materials with one or more surfactants. The polymeric material is often other than the plastics to be colored. For example, ethylene-vinyl-acetate and low-density polyethylenes are often used as dispersion media for predispersed colors used in olefinic resins because of the better dispersion characteristics of the former, capacity for high pigment loadings (thus low resin dilution) and compatibility with a variety of other resins. Low melting polyamide resins having excellent pigment dispersion characteristics are offered for the same purpose. For solution vinyl coatings, frequent use is made of a copolymer resin under conditions that provide high shear.

Pigment disperisons are generally marketed for use in specific plastics or, at the most, for a group of closely related plastics. Some so-called universal dispersions are also offered to the plastics industry, but a truly universal dispersion has not yet been produced.

In general, easily dispersible pigments are significantly more expensive to buy than their conventional equivalents. The major savings that the pigment manufacturer can offer the consumer come from more efficient usage and easier processing potentialities, which result in improved technical quality such as lower speck levels, improved brightness and higher consistency between batches. These are all desirable traits but difficult to measure in terms of costs; on the other hand, savings from increased output, more economic use of plant and labor, and improved cleanliness in handling can be easily assessed. Large savings result from the use of the more modern "stir-in," or continuous, methods of production instead of the older batchwise conventional milling operations.

Although the definitions are confusing, a color concentrate is generally understood to be color dispersion containing one or more individual colorants in the same plastic medium in which it is to be used. The colorants are provided in pellet form of desired size. The pigment loading is generally in the range of 10 to 20%. Color concentrates are designed to give a particular color when letdown at a specified weight ratio of virgin resin. Letdown ratios of 1:10 to 1:20 are common. It is estimated that about 80% of the colorants presently used in the plastics molding industry are in the form of concentrates. The use of color concentrates represents a convenient way of in-plant coloration of plastics, although its use may limit resin selection, color matching, and economics. These factors are the topics of the next Chapter.

BIBLIOGRAPHY

1. Reeve, T. B. and Zabel, R. H. "Pigment Dispersion in Plastisol and Calendered Vinyls." *Color Eng.* **3** (6) (Dec. 1965).
2. Huckle, W. G. and Jones, G. F. "Use of Pigments in Plastics." SPE Annual Technical Conference **XX**. Atlantic City (1966).
3. Miller, D. P. "Pigment Dispersions." *Color Eng.* (July/Aug. 1968).
4. Wiest, W. D. "Dissolving Pigmented Vinyl Chips." *Color Eng.* **I**, (2) 43–44 (1969).
5. Reeve, T. B. and Dills, W. L. "Principles of Pigments Dispersion in Plastics." *J. Color Appearance* **I** (1), 25–29 (1971).
6. Deverell-Smith, R. "Easily Dispersed Pigments." *Polym. Age* **2**, 267 (1971).
7. Berenson, M. "Dispersion Problems – Magnified." SPE Annual Technical Conference **XXX**. Chicago (1972).
8. Smith, M. J. "Some Physical Properties of Pigments – II." *J. Oil Colour Chem. Assoc.* **56**, 156 (1973).
9. Smith, M. J. "Some Physical Properties of Pigments – III." *J. Oil Colour Chem. Assoc.* **56** (4), 165 (1973).
10. Mess, W. H. and Garrett, M. D. "The Dispersion of Carbon Black." *J. Oil Colour Chem. Assoc.* **54**, 24–83 (1971).
11. Hackney, H. U. "Liquid Colorants for Thermoplastics." SPE Regional Technical Conference, Cherry Hill, New Jersey (Sept. 1973).
12. McCormack, T. A. "Dispersions for the Plastics Industry." SPE Regional Technical Conference, Cherry Hill, New Jersey (Sept. 1973).
13. Topcik, B. "Carbon Black Dispersion and Screen Pack Plugging in Crosslinked PE." *Mod. Plast.* **94** (Apr. 1974).
14. Smith, M. J. "Some Physical Properties of Pigments and Their Importance in Relation to Dispersion in Plastics." *J. Oil Colour Chem. Assoc.* **57**, 36–43 (1974).
15. Hackney, H. U. "Reconsider Liquid Coloring for Plastics." *Plast. Eng.*, 61–63 (March 1974).
16. Deverell-Smith, R. "Easily Dispersed Pigments." *Polym. Age* **2**, 267 (1971).
17. Meguro, K. "Changes in Surface Properties of Pigment Mixtures on Milling." Farbe and Lacke, No. 6, pp. 503–508 (1973).
18. Hafner, O. "Light Fastness of Organic Pigments as a Function of their Particle Size Distribution." *J. Paint Technol.* **47**, 9 (Oct. 1975).
19. Nanfeld, R. E. "A Method of Measuring Color Development of Color Concentrates." SPE Annual Technical Conference (1977).

7 Color Matching and Control

INTRODUCTION

Simply stated, color matching implies

Color A = Color B,

although in actual practice, the statement that

Appearance A = Appearance B

is more accurate. The problem in color matching is: given color A, how are the colorant's variables (type of pigment, degree of dispersion, quantity, ratios, etc.) adjusted to reproduce this visual response? This is a very complicated problem for which no universally accepted solution can be given. A good color matcher is the product of extensive training, long experience and, above all, good intuition. This chapter deals with some of the aspects of this work and offers a guideline to the novice.

As useful as chemical classification of colorants may be for understanding and predicting performance, the usual problem facing a formulator is that the color he must match has already been predetermined. He is faced with selection of the right colorants in the right amounts to produce that shade and offer the properties desired – all at economical costs. The reader may have formed the idea from Chapter 3 that the colorist has a large pallet of pigments and dyes at his disposal for various end uses. In fact, because of the chemical nature of some plastics and the various requirements imposed by the end use, the colorist generally has a limited number of colorants from which he can select to produce a particular shade. After meeting or compromising with the requirements discussed in Chapter 5, the pallet becomes rather limited, and the choice of pigments does not remain as broad as it appears to be initially.

Once the colorist has critically examined the available colorants for color, strength, cost, dispersion, performance, process stability, compatibility, etc., he has to decide which colorants will provide the required match. This depends on the following:

1. The sort of color to be matched. Is it opaque, transparent, or translucent, and does it have any special features, such as the presence of fluorescent or metallic pigments?
2. If it is transparent, can the requirements of heat and light stability be met with dyes or must there be a search among the pigments for those that can give transparent effects?
3. If it is opaque (pigments will undoubtedly be used), are there pigments in stock which are bright enough to achieve the required color?
4. If a translucent color is needed, what additional requirements are there? Will the color have to be similar when viewed in reflected and in tramsmitted light? What sort of illumination will be used? If it is illuminated from behind, does the color have to show sufficient diffusion to obliterate the outline of the lamps or even the lamp filaments themselves? The answers to all of these supplementary questions will determine whether the translucent color will be based on pigments or on a mixture of dyes and a filler that provides light-diffusing properties.
5. How metameric will the matching be? Will metamerism be an important consideration in this instance, and can anything be done about it anyway? If something can be done, is the sample large enough to permit measurements to be made of its spectral transmission or reflection?
6. How accurate must the match be and how consistent must successive production runs be?

Answers to all of these questions are difficult to obtain initially, but they emerge as the colorist attempts to obtain customer approval for the matching.

Selection of the proper colorants is of major importance, but even that decision does not provide information on the correct *amount* of colorant to use. This is traditionally determined by trial and error, relying on the skill of the visual color matcher, but more and more often, instrumental and computational methods are supplementing this skill.

Regardless of whether visual or instrumental methods for color matching are followed, the color formulator always starts with a sample of the color the marketing department or the stylist wishes him to reproduce. This may be a piece of fabric, a painted chip, a plastic article, or even a particular area in a painting or a colored photograph. Two questions need to be asked immediately. The first is whether samples are required to look alike under all light sources. The second is how close the match should be; in other words, what the tolerances will be. The answers to these questions will save the colorist's time and avoid later arguments with the marketing department or the customer over whether the submitted match really is a match.

Color formulating activity can be divided into two parts: 1) The preparation of an initial match, which in practice includes the selection of pigments; and 2) the adjustment of a previously formulated match to conform to a scale-up production. This is generally called tinting, shading, or simply color adjustment.

In addition to the above, the plastic colorist is charged with the responsibility of maintaining the uniformity of the color in the product. The techniques and objectives of initial formulation and adjustment and of quality control differ somewhat, but the general principles are the same.

VISUAL COLOR MATCHING

An experienced formulator usually starts with a reasonably good idea of those colorants he needs in order to match the color. He also knows that he will probably have to undergo considerable trial and error before he finds the exact combination of colorants for his purpose. Frequently, many laboratory or pilot plant batches must be made before the right combination of pigments is established. Visual color matching is evidently an involved and time-consuming method. The task becomes more difficult when a spectral match is required.

Fine adjustment of shade is often more complex than the original approximate color matching. The laboratory batch may have the correct pigments but not the correct amount, and will then have to be shaded with more of the same colors until the desired color is obtained. In some cases, the color may have to be adjusted by adding other colors. Such additions can affect other characteristics required in the match. When the laboratory batch is satisfactory for color and all other characteristics, a production-batch formula is written.

With some reservations, one could say that the foregoing procedure is a trial-and-error method and not very scientific. Yet, this has been the standard practice in the plastics industry and is still used to a great extent. While it is fairly simple, it is extremely time-consuming. Hence, more efficient methods are required.

INSTRUMENTAL COLOR MATCHING

Instrumental methods of color matching are essentially an extension of visual color matching. The functions of the human eye and brain are taken over by a spectrophotometer and other color-measuring instruments, discussed in Chapter 2, and the trial-and-error process of color proportioning is performed by a computer with no actual mixing of the colorants. The techniques of instrumental color matching have been developed primarily for paints and textile industries. The purpose of the present discussion is to point the reader to the prospects of using these techniques in the plastics industry and inform him of the limitations he may face in employing instruments for his color-matching function.

Spectrophotometers, along with various attachments (log-density cam and R-density cam) have been used extensively to detect choice.

All the systems on the market are basically the same in that they use a spectrophotometer to measure reflectance values, an input/output device (usually a teletype), either a dedicated or time share computer, and associated programs or software. The operator measures the reflectance values or his standard and selects the pigments to be used to make a match. The system then matches to the reflectance values of the standard by adjusting the relative concentrations of the pigments that were selected for the match. If the operator has made a wise selection of pigments, the system will return a formula in terms of relative percent concentration, which must then be converted into amounts to be weighed by selecting a percent loading.

Spectrophotometers can easily be used to achieve invariant color match, using the same colorants as those present in the sumitted sample. However, it is generally not possible to use the same colorants as those present in the submitted sample, which might be a swatch of cloth, a sheet of another plastic, a chip of paint, etc. In such a case, one must work with color coordinates of the sample – for example, those related to CIE, or Hunter *L, a, b* systems. As discussed in Chapter 2, these colorimetric techniques are based on the principle of substractive mixing and use approximations of Kubelka-Munk-Judd equation. Under the assumptions and restrictions for the applicability of this equation to the system, one can consider

$$\text{Color A } (X, Y, Z)_a = \text{Color B } (X, Y, Z)_b$$

as the basic relationship for instrumental color-matching techniques.

Since the color as seen by the observer is an integrated result of the interaction of the light source, reflectance, and observer's sensitivity, there could be many combinations to fulfill the basic equation $(X, Y, Z)_a = (X, Y, Z)_b$. In general, these combinations are apt to be quite sensitive to the nature of the light source and to the observer. Such matches are called *variant* or *metameric* matches.

A number of analytical and empirical relations are available which can be used for the selection of suitable colorants and establish their respective concentrations for any particular match. The solution of these relations needs extensive computation. COMIC (Color Mixture Computer) has been widely used for this purpose but is being replaced by digital computers. Computers permit the entire process of color formulation to be carried out in 15 to 20 min. The first formulation is generally a trial, not a match of the material on hand. Further refinements introduce the difference in tristimulus values and visual inspection of the samples.

The basic data for the computations are provided in the form of reflectance curves of the individual colorants and of the standard match. A necessary preliminary step toward successful computer formulation is, therefore, the development of a colorant library. This consists essentially of data derived from a series of spectrographic curves on each of the colorants normally available to the formulator. The colorants are dispersed in the plastic in 10 to 15 levels of concentration, ranging from 0.05 to 2%. As the nature of plastic markedly affects the hue and chroma of the product, a separate color library is required for each type of plastic, or at least for each class of plastics, such as polyolefins, nylons, acrylates, etc. These data are referred to as pigment data.

The pigment data must be qualified with the type of stabilization formulation and the presence or absence of any delusterant in the formulation. The first qualification is necessary because some stabilizers have their own color which they impart to the plastic. The second qualification is necessary because color instruments cannot distinquish – as the human eye can – between translucent and opaque colors: in a spectroscopic sense, white is not a color.

A given color can be prepared by using several formulas consisting of different pigments. These formulas may differ in their performance characteristics as well as in cost. The approximate formulas, derived from the color-matching instrument and computer calculations, are used to determine whether, for example, the small amount of the expensive organic pigments or the large amount of the cheaper inorganic pigments is more economical. Herein lies probably the strongest incentive for employing color-measurement instruments for color matching.

In addition to the computer generated formula, the system is capable of making the following calculations:

1. A predicted color difference or DE is given for three basic types of lighting conditions: daylight, tungsten, and fluorescent. This difference is a measure of the degree of metamerism.

2. The color correction program generates a corrected formula or second trial by measuring the reflectance values of the standard and first trial. These values, along with the pigments and amounts used in the first trial, are entered into the computer. It adjusts the relative percent concentrations and comes back with a corrected formula for a second trial.

3. The capability of manual adjustment of concentrations allows the formulator to manually manipulate or specify the relative percent concentrations of pigments in a formula. It can be used to generate reflectance values of a theoretical color by selecting a number of pigments and then specifying to the computer relative percent concentrations for each pigment. In a case where the formulator has an off-shade pigment in his inventory that he wants to work off by using it as a partial replacement for the on-shade pigment, the degree of mismatch between the two formulas can be predetermined by generating theoretical reflectance values for each formula and then asking for a color difference.

4. The chromaticity coordinates *(x, y, Y)* are calculated for the standard and trial every time the color match or color correction programs are run. In addition, the coordinates are generated when the program for manual adjustment of concentration is used. Plotting the chromaticity coordinates of the standard and trial of a troublesome match will indicate whether or not it is even possible to obtain a match with the pigments available to the formulator. These coordinates are also very helpful when selecting a candidate pigment for replacement. This point is elaborated in some detail in the following, as the colorist is very often confronted with this problem due to governmental regulations (e.g., replacing lead containing pigments) or due to nonavailability of some pigment.

Using the feature of manual adjustment of concentrations the CIE curve of any pigment in tint form can be generated and plotted on the CIE chromaticity diagram. If this is done for the standard pigment and the candidate, a colorist comparison can be made. The procedure is to generate x, y, and Y for theoretical mixtures of the colored pigment with white. The chromaticity coordinates, x and y, for each tint in the sequence are then plotted on the CIE diagram. The candidate can then be judged as to how good a replacement it would be in terms of hue and chroma to the standard, from very light tints all the way to the mass-tone level. In addition, an indication of strength can be interpreted from the two curves by observing the position of each point relative to the neutral axis and the distance between each successive point on the curve: the farther away from the neutral axis, the stronger the pigment.

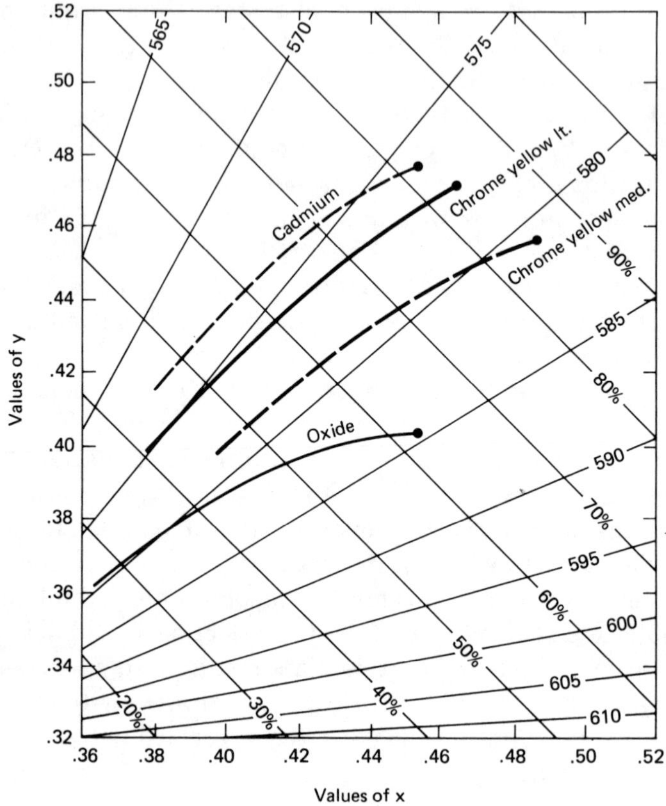

Fig. 9. Chromaticity diagram for four yellow pigments. Source: Osmer, D. *SPE ANTEC* (1977).

Figure 9 shows a cadmium lemon yellow and a yellow oxide pigment relative to two chrome yellows. Clearly, chrome yellow light is greener than chrome yellow medium. The relative straight line travel of these plots is indicative of high scattering pigments (i.e., in combination with TiO_2 there is essentially constant scatter).

The cadmium is slightly greener than the chrome yellow light at all levels. If the cadmium was an acceptable replacement, it could be combined with a red shade yellow to match both the hue and chroma of the chrome yellow light. The yellow oxide is much redder and lower in purity, and is not, by itself, a good candidate for chrome replacement. However, the relative straight line behavior indicates high scattering. Therefore, this colorant might be an excellent candidate if used in combination with a high chroma colorant.

It should be emphasized that a final selection must consider more than a match of the color, i.e., its hue, saturation, and lightness. These plots are only one of many tools to be used. Heavy metal content, lightfastness, migration, chemical resistance, cost, and chemical composition must also be considered.

Two diazo pigments and two isoindolinone pigments which are available for chrome yellow replacement are shown in Fig. 10. These four pigments are more transparent than the chrome yellows, as indicated by their curved lines.

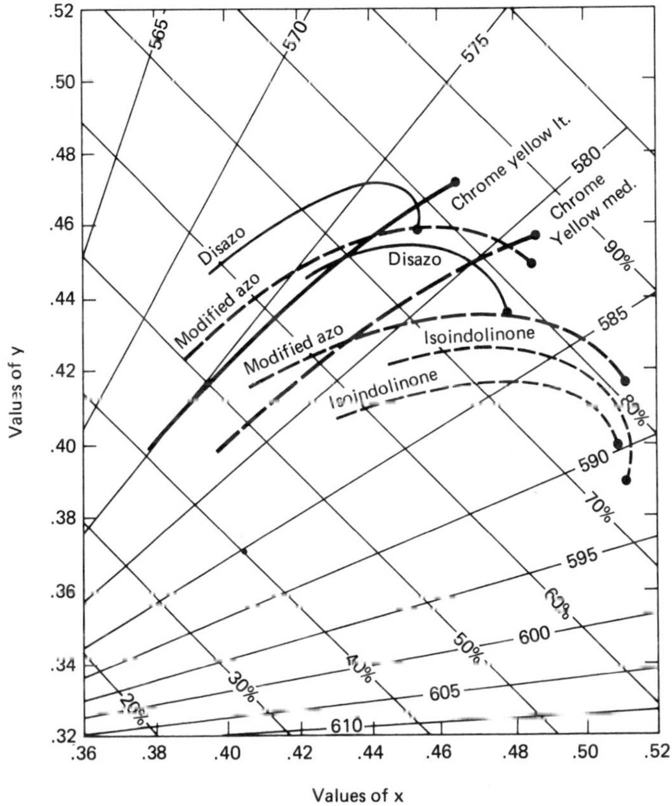

Fig. 10. Chromaticity diagrams of two diazo and two isoindolinone pigments, which can be considered as replacements for chrome yellows. Source: Osmer, D. *SPE ANTEC* (1977).

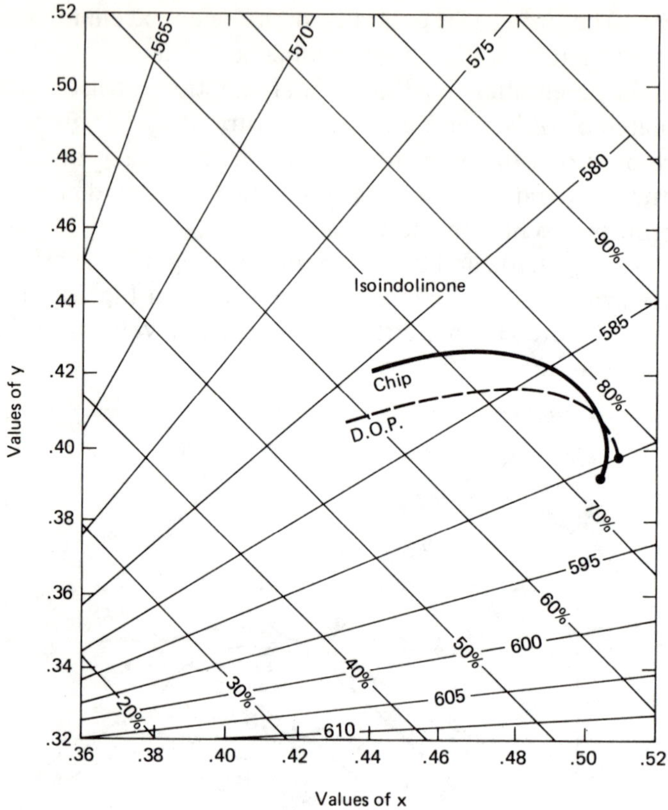

Fig. 11. Cromaticity diagram of isoindolinone pigment as D. O. P. and chip dispersions. Source: Osmer, D. *SPE ANTEC* (1977).

In general, the more nearly the plot follows a given dominant wavelength, regardless of letdown level, the easier it is for an inexperienced visual matcher to arrive at a match. Beginning visual colormatchers frequently have difficulty in understanding that when white is added to make a color lighter, it may also shift the hue (e.g., yellows tend to shift slightly greener). This phenomenon seems to be more prevalent with organic pigments.

Although no colorant is an exact replacement for the chrome yellow pigments, the redder shade diazo (Pigment Yellow #95) would be an appropriate consideration for replacement of chrome yellow medium.

Two modified azo pigments, which would be appropriate considerations for replacement of chrome yellow, are shown in Fig. 10. As can be seen on the graph, the modified azo pigments do a good job of bracketing the chrome yellows. Although chrome yellow still cannot be matched in the masstone area, these new pigments are more saturated than some of the other organic pigments.

It must be cautioned that the plotted chromaticity properties of pigments discussed here are for the colorants in flexible PVC using DOP dispersions, and that the samples were at complete opacity. Deviations from these conditions would change the relationship. To illustrate, a plot of an isoindolinone pigment is included showing both DOP and vinyl chip dispersions, but still using flexible PVC. As can be seen in Fig. 11 the vinyl chip dispersion is more transparent, achieves greater saturation, and covers a wider hue range than the DOP dispersion. Most pigments, in fact, behave similarly.

LIMITATIONS OF INSTRUMENTAL COLOR MATCHING

Instrumental color matching has been the center of colorists' attention for the last several years. Indeed, much progress has been made and developments are continuously being reported. Unfortunately, however, there are numerous limitations which constitute a barrier to realizing the goal of color matching in plastics solely by reference to instrumental methods. The plastics industry shares some of these limitations and deficiencies with colorimetry in the paint and textile industries, but it also has its own special problems.

First of all, the measuring instruments themselves have a major defect in their application to plastics. This is the small size of the sample viewing area. A visual assessment taken over a large area automatically integrates the small color variations over the surface of a sample. When these differences exist, as they do frequently on darker colors, the general effect is what should be measured. This would require an instrument with a large viewing port, but with precision equal to instruments with a small port.

A second limitation to the instrumental measurement of color in plastics is that the sample *K/S* ratio function of the Kubelka-Munk-Judd equation does not work well for plastics. The use of (K/S) 1.25 reportedly permits well-balanced first predictions, i.e., predictions about the required hue. A subsequent tristimulus difference correction easily corrects chroma and value. Added to the difficulties of applying the Kubelka-Munk-Judd equation to opaque plastics is the fundamental nonapplicability of this equation to transparent and translucent media. Thus, the matching and control of translucent colors in plastics is inherently difficult, if not impossible, to accomplish by instrumental color-measuring techniques. Plastics share this problem with glass.

The third difficulty with color-measuring instruments is that they cannot give a reliable measure of reflectance of the standard. The failure is caused by lack of correspondence of instrument geometry with visual viewing conditions and insensitivity of the instrument.

Both instrumental and visual matching of fluorescent materials can be difficult. Fluorescense presents a new variable and introduces the possibility of a fluorescent metamerism where two samples match under one source, not under another, more active source. The same applies to metallic and to iridescent colors, where not only the source is important but also the angle of the incident light as compared with the viewing positions.

The purpose of instrumentation in color matching is to reduce the amount of experience required on the part of the colorist. In the last analysis, however, the colorist must acquire some experience in the use of instruments and in the interpretation of their results before he can find them useful aids in the color-matching process. In the coloring of plastics, where there is as much art as science, the experience of the color matcher must never be underestimated.

Considering the complexity of color matching, it is unlikely that the problem will ever be handled entirely by instruments. Rather, the color matcher will combine his expertise with the use of spectrophotometry, colorimetry, and computer methods. These aids will relieve him of much of the routine, sample-matching tasks and enable him to cope more intelligently with the task.

COLOR ADJUSTMENT

The initial color match generally does not correspond with that actually obtained in production. In all probability the production match will be close, but not close enough to be the initial pilot plant or laboratory match. The reasons for the differences include color-measurement errors, batch-to-batch variation in pigments and in virgin plastics, color-mixing anomalies, and the machine factor in extrusion and mixing. An adjustment is therefore often required. For the visual color matcher the process of adjustment is identical to that of producing the initial match. He continues the process of repeated trials until the match is achieved. For computer techniques the adjustment procedure involves calculating corrections to the concentrations used in the trial sample. The corrections required are those which will make the sample's color coordinates the same as those of the standard.

Another problem in the coloration of plastics is the change of chroma and hue experienced with pigment development as it is dispersed in the medium. This change appears to be related to pigment particle size, organic pigments being more changeable than inorganic pigments. The only solution is a precise correlation between laboratory and production scale units. There is no short cut, particularly with processes that use ball mills to incorporate pigments, because color development is continuous until primary particle size is achieved and particle distribution is uniform. Since the development is not linear with time, the time factor is of obvious importance if the color is to be "caught as it runs," and it is advantageous to work on a fairly flat part of the development curve if at all possible.

VARIANT AND INVARIANT MATCHES

It has been pointed out that the physical stimulus which the brain converts into our concept of color is made up by combining, wavelength by wavelength, the spectral reflectance curve of the object, and the spectral response curve of the detector. It follows, therefore, that the stimuli to the brain and, along with it, the perception of color, will differ if any of the variables change. Thus, two objects seen under one illuminant (daylight) may match, but when seen under another illuminant (incandescent lamp or fluorescent tube) may not match. Similarly, two objects may appear to be a perfect match to one observer but it may be a poor match to another observer. Such objects are called metameric objects or a metameric pair. They are also referred to as a conditional or variant match. While the observer's metamerism is not very common, the illuminant metamerism is a constant problem.

The requirement that the two objects look alike under all light sources means that they must have identical spectral reflectance curves. This type of match is called nonmetameric or *invariant match.* An invariant match requires that identical colorants be used in the sample to be matched. It is, therefore, quite unlikely that a plastic color could be formulated to be an invariant match, say, to a dyed fabric or a paint chip. To achieve an invariant match is difficult even in cases where one plastic material is to be matched with another plastic material. There are several reasons for this: the degree of dispersion of the same pigment differs from plastic to plastic; a shading or tinting may be required in one plastic to overcome the yellow or greenish hue of the virgin polymer, while for another plastic no such adjustment may be required; and the mode of fabrication may differ in the two cases, and this may make the machine factor important for the match.

Metamerism is a major source of customer complaints and rejected material. Therefore, even in cases where the achievement of an invariant match is not possible, attempts should be made to reduce metamerism as much as possible. If both spectrophotometric and colorimetric instruments are available, the initial color formulation should be done on a spectrophotometric basis and corrections made on a tristimulus basis. The achievement of a nonmetameric match without spectrophotometer is even more difficult. But there are some rules of coloring mixing which, if followed, can be of great help in reducing metamerism. These rules have been excellently described by Longley and Winey (see Bibliography) and can be summarized as follows:

1. The closer together the yellow and orange (or red), the greener the match will be in tungsten; the farther apart they are, the redder the match will be in tungsten.
2. When two combinations in the brown region match in daylight and tungsten, the one with the greenest shade of yellow will be red in fluorescent.
3. With any green or blue; the greener the yellow, the greener will be the color in tungsten. Conversely, the redder the yellow, the more clean and green will be the match in tungsten.

4. With a given yellow, the redder the shade of blue, the greener will be the color in tungsten. Or, the greener the shade of blue or green, the cleaner will be the color in tungsten.

After choosing pigmentation for a daylight-tungsten match, the colorist may find that the color difference is too great in fluorescent. Adjustment for the third lamp requires keeping in mind what shades were used – a relatively greenish or reddish shade of yellow or a relatively greenish or reddish shade of blue. Starting from a color consisting of the two pigments of each hue group whigh are closest in color, it is possible to stay reasonably close in daylight and tungsten by moving to a redder yellow while also moving toward a redder blue. A redder yellow, taken alone, would tend to go clean in tungsten; a redder blue would nullify this because, taken alone, it would tend to go grey in tungsten.

This leads to Longley and Winey's Rule 5 which says that opening the gap in hue between the two groups leads to a relatively cleaner shade in fluorescent; closing the gap in hue difference makes the color go greyer in fluorescent. This, of course, also requires that the choice of pigments be shifted appropriately in each direction so as not to unbalance the daylight-tungsten match.

Some compromises are inevitable because the colorist's choice of pigments is limited. But if he wishes, he may control the color in three illuminants by using a blend of green- and red-shade blue for blueness and a blend of primrose and medium for yellowness. This results in four pigments, normally undesirable but a useful device since a spectrophotometric match will not be likely.

When close tolerances are not essential, a painted patch or a strip of colored textile is a perfectly adequate sample for matching, even though the plastics colorist's sample will almost inevitably be a metameric match to the original because of the different colorants employed. However, a painted or dyed sample is a far less satisfactory sample when it is used by two suppliers to the same customer, because the metamerism will lead each to interpret the color differently and their respective matchings may be unacceptably different. The painted or dyed original is unacceptable when it is to be used to prepare a number of articles in different plastics for assembly of an appliance that must present a uniform appearance in all lights and that must be sufficiently consistent in successive production runs for replacement parts to match. When this demand arises color standards and formulations with mandatory pigment type are necessary to achieve perfection.

The preceding discussion is not to be construed as suggesting that metamerism should be avoided at all costs. However, one should be fully aware of the dangers involved when deliberately producing metameric matches. It should be clearly understood by the supplier and the customer that the match is conditional. And when it is agreed that a conditional match is adequate, it is important to know under what illuminant the match will be judged.

DICHROISM

Dichroism is a property of transparent colorants in which the hue may change with colorant concentration or layer thickness. Dichroism is an instrinsic property of a transparent colorant, dependent upon the shape of its transmission curve. If the curve is symmetrical there will be no dichroism. Yellow, orange, red, and violet colorants show dichroism in decreasing order because of the dissymmetry of their spectral transmission curves. Conversely, blue and green colorants tend to have symmetrical curves and show little or no dichroism. Dichroism creates difficulties in the color-matching process. When the concentration of a colorant is altered, the color of the object does not simply become more or less intense, it usually also undergoes a hue shift. The phenomenon of dichroism is widely known in the coloring of transparent plastics and is also occasionally encountered in translucent or even in some opaque plastics.

Two rules are suggested for dealing with dichroic colorants:

1. With increasing colorant concentration, the areas under the lower peaks of the color spectrophotometric reflectance curve decrease proportionally more rapidly than the areas under the higher peaks of the reflectance curve until a certain minimum is reached, resulting in a hue shift toward the dominant wavelength of the area under the highest peak.

2. If there is only one reflectance peak in the color spectrophotometric curve, with increasing colorant concentration the areas

under the lower parts of the reflectance curve decrease proportionally more rapidly than the areas under the higher parts of the reflectance curve until a certain minimum is reached, resulting in a hue shift toward the dominant wavelength of the peak of the reflectance curve.

Many of the dyes used to color transparent plastics are also soluble in solvents. A simple way to test for dichroism is to prepare solutions at several concentrations and to observe any hue differences among them. This assumes that the dye will have the same color in both the solvent and plastics, which normally is so.

THERMOCHROMIC EFFECT

The color of some pigments is changed by changes in temperature. This phenomenon causes severe mismatches of finished products which seemed well-matched at the time of production. The most striking instance is that of certain pyrazalone reds, particularly in a tint with titanium dioxide, which change from their normal pink color to a much more orange hue at processing temperatures. Unfortunately for the color matcher, the hue reverts to pink only after up to 24 hr. at room temperature. As a result, freshly prepared plastic showing a perfect color match becomes much more reddish after it returns to its equilibrium color This is probably due to different phases or crystal forms being stable at the two temperature ranges, equilibrium being reached slowly.

Some other pigments also show this thermochromic behavior, though to a lesser extent. The use of severely thermochromic pigments should be avoided. If they must be used, color should be compared with a master swatch heated to processing temperature or adjusted to give an instrument reading corresponding to this higher temperature. Obvious precautions analogous to those stated for metameric matches should be observed, such as not using a thermochromic red in combination with a nonthermochromic red.

COLOR CONTROL

Once the customer has accepted a color formulation and production is started, it is necessary to *control* the process variables to maintain and deliver the desired color. Inherent in the decision of whether the process is in control is the need for establishing tolerances. This is particularly important when plastics from different production runs are later combined in one end product, such as the sleeves and body of an outerwear made of PVC leather fabrics, automotive trimming, appliances, etc. Often a color standard must be maintained year after year, for example, in telephone sets, wire coating, traffic signs, car taillights, etc.

TOLERANCES

Since it is not often economically feasible to match color standards perfectly in production, the limits of permissible color deviation should be included in every specification. This deviation, or tolerance, must be small enough to ensure that the color is satisfactory, yet it must be broad enough to permit economical manufacture of the product. An all-too-common system of establishing color tolerance is to state that the batch must be a "good visual match to the standard." In effect, this is no guidance at all since it leaves everything to personal opinion. Such opinion is influenced by many extraneous factors, including the end use of the product. With no guideline other than the admonition to provide a "good" match, limits might be interpreted too closely or too broadly.

In view of this, it is advisable to set limits related to the three dimensions of color: hue, chroma, and lightness (or value). Sometimes limits for only one or two of the dimensions are given. This tolerance specifies the hue difference in terms of green and red variations from the standard. No limits are set for saturation or lightness, so these dimensions are a matter for personal opinion or arbitration. If limits were set on these attributes of color also, the observer would have a much more complete idea of what is required.

At present, the plastics industry uses three methods for determining whether a production color is within the tolerance permitted: visual examination, instrumentation, and a combination of the two. For visual examination, two and preferably three standards should be provided which differ within the tolerance limits in hue, chroma, and saturation. Sometimes such a provision is not possible. In these cases, a *grey scale* tolerance is specified. The use of the grey scale method of evaluating color differences would seem to indicate the feasibility of equating a lightness difference with a chromaticity difference. For small differences this practice produces consistent results; however, the relation is not constant and fails in large color differences.

Color-measuring instruments are increasingly being used to specify colors and impose tolerances. Here, too, the three dimensions of color have to be specified for better understanding of the color tolerances. Some equations have been proposed which combine the three dimensions into a single number, but these color-difference equations have provided limited utility for plastics. No simple, constant ratio between the lightness difference and the chromaticity seems to produce generally satisfactory results.

COLOR DRIFT

Color drift is a collective term for many problems which produce color changes from run to run or during one production run. A successful quality program requires control of all these factors and recognition of their interplay. For systematic treatment, they may be divided into problems of matching and viewing, on one hand, and control of actual color changes during the production cycle, on the other hand. The problems of the first category have already been discussed in this Chapter and elsewhere in this book. The problems of color control during production are related to the control of raw materials (colorants and plastics) as well as to the control of processing conditions.

Incoming pigments, pigment dispersions and color concentrates must be checked against the specifications, which generally include average particle size, moisture content, and specific gravity. In case of color concentrates, melt viscosity is also checked. The tinctorial characteristics should be checked against a standard pigment or dispersion originally provided by the supplier. Usually a laboratory swatch of the pigment shipment is run in white let-down and the resulting spectral curve is compared to the standard curve. If spectrophotometric equipment is not available, a visual comparison should be made under standard light. This provides a qualitative approach to acceptance decisions.

The final color of the product is dependent on the coloring characteristics of the pigment and the inherent color of the polymer; therefore, it is imperative that the color of plastic raw material, whether virgin or reground, be kept within specification. The judgments made on the appearance of granules, chips or powder are often misleading. A lab swatch should be run and light transmission measured photospectrometrically or visually. Care should be taken in sampling the shipment. All additives, except the colorant, should be added to prepare the test sample.

The processing problems which affect the color uniformity are varied. The errors in calculations, weighing and blending, and contamination with other colorants or color-producing impurities are obvious. Many pigments are affected by high temperatures: the close control of extrusion on die temperature, therefore, becomes necessary.

Many instances of color change are caused by yellowing or darkening of the resin matrix rather than failure of the pigments. This factor becomes most visible in plastics of light color and low pigment content. Steps used to alleviate it include decreasing the heat exposure time or processing temperature and careful design of the stabilizer system. No one stabilizer meets the needs of every resin or every processing cycle. Optimizing stabilization generally eliminates discoloration of the plastic without sacrificing a workable temperature schedule.

A much more common cause of color drift during production is a gradual increase in the degree of pigment dispersion, especially when dry pigments of difficult dispersion characteristics are used rather than a predispersion. A dry pigment, incompletely dispersed, may yield a sufficient portion of its ultimate color strength to offer apparent economies, but it undergoes further dispersion, and thereby develops higher tinctorial strength, as the plastic is subjected to shear during processing. By using pigments predispersed to ultimate particle size, changing color development is prevented. As the pigment dispersion is affected by back pressure of the extrusion screw, the control of this parameter is helpful in avoiding any possible color drift.

A special aspect of a shear-related problem occurs in the use of reflective flake pigments, either metallic or pearlescent. While these flake pigments are much more easily dispersed to their true particle size than are colloidal pigments, the individual flakes themselves are often fractured by mechanical strain and shear. This results in lower brilliance and increased opacity; in particular, the overall color of a mixture of metallics with colloidal color pigments is affected severely. These problems are largely eliminated by selecting a grade of reflective pigment suitable for the particular application. Modifying the plastic processing cycle in order to avoid excessive shear during vulnerable stages of processing of the flake is also beneficial. Predispersion or prewetting of the flake pigment frequently avoids initial grinding of flakes against each other at high metal concentration, which contributes to breakdown. Adding the flake pigment after the resin and auxiliary additives are fluxed to a plastic state avoids subjecting the pigment to the extreme shear prevalent during fluxing of the solid resin.

Color audits should be conducted periodically. Retained samples of the finished goods, coated fabrics, films or extruded parts should be visually checked against the master on the spectrophotometer. Color-difference values should be viewed and, if necessary, steps taken to improve color formulations.

To sum up, prevention of color drift problems must start with an easily controllable color match, continued through a carefully engineered production cycle, and end successfully with final quality control and inspection.

BIBLIOGRAPHY

1. Kiernan, W. J. "Color Specifications for Telephones." SPE Annual Technical Conference **XX**. Atlantic City (1964).
2. Johnston, R. "Selection and Training of Color Matches from a Computer Color Control Program." *Color Eng.* **3**, (6), 20–21 (1965).
3. Davidson, H. R. "Instrumental Matching and Control of Metallized Colors." *Color Eng.* **3** (1), 22 (Jan./Feb. 1965).
4. Davidson, H. R. and Hemmendinger, H. "Use of Instrumentation in Color Matching Problems." *SPE J.* **19**, 104–108 (1963).
5. Macbeth, N., Jr. "Color Matching." *Illum. Eng.* **59**, 461–471 (1964).
6. Longley, W. V. and Winey, R. K. "Practice Solutions to Problems of Metamerism." SPE Annual Technical Conference, Detroit (1967).
7. Webber, T. G. "Dichroism of Transparent Colorants." *SPE J.*, 29 (Sept. 1968).

8. Felsher, H. and Hanan, W. J. "Avoiding Color Drift in Vinyl Processing." SPE Annual Technical Conference **XIV**, New York (1968).
9. Huey, S. J. "Standard Practices for Visual Examination of Small Color Differences." *J. Color Appearance* **I** (4) 24–26 (1972).
10. Huey, S. J. "Scientific Approach to Visual Color Comparison." *J. Paint Technol.,* **44**, 83–89 (1972).
11. D' Amico, J. N. "A Study of the Phenomenon of Dichroism in Vinyl Plastics." SPE Annual Technical Conference, New York (1970).
12. Mertz, C. H. "The Applications of Mini-Computers to Industrial Color Control Problems." SPE Regional Technical Conference, Cherry Hill, New Jersey (Sept. 1973).
13. Venable, W., Jr. "Specification and Measurement of Optical Properties of Translucent Materials." SPE Regional Technical Conference, Cherry Hill, New Jersey (Sept. 1973).
14. Best, R. A. "Color Formulation and Control in Plastics Industry." pp. 128–145 in R. M. Johnston and M. Saltzman, *Industrial Color Technology,* American Chemical Society, Washington, D. C. (1972).
15. Johnston, R. M. "Color Control – The Art That is a Science." *Plast. World* **62** (July 15, 1974).
16. Holtzen, D. A. "Spectrophotometric Matching of Opaque Plaques." SPE Annual Technical Conference (1975).
17. Bowman, J. F. "Replacing Lead Containing Pigments in the Computer Color Matching." SPE Annual Technical Conference (1977).
18. Osmer, D. "Using Chromaticity Diagrams for Pigment Selection." SPE Annual Technical Conference p. 38 (Sept. 1977).

8
Compounding

INTRODUCTION

Adding and mixing colorants into plastics is an integral part of a much broader operation called compounding. Few resins are useful in their natural form; they are, therefore, mixed with other materials to improve or enhance their properties, making them more useful for a variety of applications. The process by which ingredients – including the colorants – are intimately mixed together into as nearly a homogeneous mass as possible is known as *compounding.* The term had its origin in the rubber industry but is now used in the plastics industry, also.

To the extent that compounding is still an art, experience and proprietary information can tilt the balance of an operation from one that is failing or marginal to one that is highly successful. Nevertheless, there is room for statements of general principles and for descriptions of fundamental processes and basic equipment. Some excellent reviews have been written on this topic: a partial list of references is given in the Bibliography. In view of these and other easily available and comprehensive treatments, the subject will be dealt with here only in outlines.

ELEMENTS OF COMPOUNDING

Because of the nature of both the resin and the additives, compounding requires a wide range of mixes (dry powders, slurries, pastes, doughy consistencies), and a corresponding range of mixing operations. The task of mixing becomes one of changing the original distribution of two or more nonrandom or segregated masses into an acceptable distribution of one mass in the other. Thus, the problem becomes one of redistributing masses in order to achieve the desired probability distribution in the absence of diffusion or other random molecular motions.

In mixing processes involving plastics the material is regarded as essentially a fluid subjected only to laminar flow. Thus, the problem of mixing in plastics is that of subjecting such materials to laminar-shear deformations in such a manner that an initially nonrandom distribution of ingredients approaches some arbitrary scale of randomness. The problem is usually complicated by the fact that the ingredients do exhibit interparticulate forces, so that the stresses accompanying the deformation must be considered, as well as the deformation process itself.

The general theory of mixing usually considers a nonrandom or segregated mass of two components and their deformantion by a laminar or shearing deformation process. The object of shearing is to mix the mass in such a way that random samples taken from the mass exhibit minimal variations. The shearing process is generalized, i.e., it is not limited to any particular kind of shearing action or mixing device, but applies to all such devices.

Three principles are given by the general theory of mixing:

1. The interfacial area between different components must be greatly increased in the mixing process because it results in a decrease in average striation thickness.

2. Elements of the interface must be distributed uniformly throughout the mass being mixed. *Uniformity* can be defined by the intensity of segregation or by the average deviation of particles from a mean concentration.

3. The mixture components must be distributed so that for any unit of volume, the ratio of components within the unit is the same as that of the whole system.

The first and second principles obviously require, for any particular mixing process or device, that the dynamics of fluid or particulate flow within the device obey the three requirements or efficient mixing: the shear should be at right angles to the direction of flow; the shear rate should be controlled by varying the gap between the stationary and rotating elements, by varying rotational speed, or both; and that each particle should follow the same flow path to experience the same shear history. They also suggest that the optimum initial orientation of the interface between components should be normal to the flow lines or streamlines within the device and that all the streamlines of flow should lead to a region of maximum shear. The last principle implies that the scale of sampling must be large in comparison to the ultimate particle size.

The mixing theory, being generalized, suffers from limitations which must be recognized when it is applied to actual plastics and mixing devices:

1. It is limited to purely viscous (Newtonian) materials, while actual plastics are viscoelastic, or otherwise non-Newtonian. However, the qualitative concepts which hold for simple viscous bodies are of great value in furthering the understanding of the mixing of real materials.
2. It is assumed that there are no van der Waals or other forces between particles.
3. Initial orientation of components is assumed to be known, although this is almost never true in practice.
4. Mixing process is assumed to be isothermal, yet the heat generated in mixing most plastics is sufficiently high so that the energy dissipation during mixing is considerable.

The major gap between theory and practice lies in our inability to fully describe the rheological behavior of complex plastic melts and plastisols, particularly since the flow within actual mixing devices is very complex. Still, the application of qualitative and semiquantitative concepts has thrown considerable light on the process of mixing.

A mixture is described in terms of the statistical deviation of a suitable number of samples from a mean, the sample sizes being dependent upon some length, volume or area characteristics of the mixture or its properties. For example, if color is imparted by adding pigment and homogeneity is measured by visual impression, the characteristic length is the resolving power of the eye, is say 0.001 in. A completely mixed compound exhibits pigment streaks no greater than 0.001 in. On the other hand, the color value for any given series of samples would appear in uniform to a spectrophotometer that integrates over a 1-in. diameter circle even if the streaks were 0.1-in. thick. Likewise, the intensity of color difference between streaks would affect the resolving power of the eye – or the spectrophotometer – and, hence, the characteristic length.

The means for measuring the degree of mixing are varied. In commercial practice, inspection for color homogeneity, color comparison for specks, streaks, or spots of unmixed filler or resin is visual. Frequently, changes in physical properties, such as tensile strength, modulus, or density, are used to evaluate the degree of mixing. Since these properties may be affected during polymer degradation and thermal effects on the mixture, property changes do not strictly evaluate mixing as such. Some of the more fundamental measures evaluate changes in rheological properties, chemical reactions, and electrical conductivity. These nearly approach the ideal measurement of the basic criteria of mixing. The only direct measure of a basic criterion of mixing is *striation thickness,* though this property is measurable at present only under certain conditions.

Because of the many variables of mixing and the scantiness of criteria for measuring mixing effectiveness, the current theory of mixing heavy viscous materials is less than satisfactory. Often, practical and extensive tests are the only means of determining the suitability of a particular mixing machine for a specified plastics mixture. Most compounding can probably be classified as an art.

The factors that contribute most to the ultimate properties of plastic compounds are many, depending on the type of basic resin, type of compound, the types and amounts of other ingredients, and the desirable degree of homogeneity. The basic resin may vary considerably chemically as to molecular weight and particle size. The compound range from an adhesive or coating solution to a molding powder or an extruded melt. In additon, there is a large number of secondary materials which may be incorporated with the plastic resin in the mixure, such as plasticizers or solvents, solids such as fillers or pigments, and others such as stabilizers, dyestuffs, and lubricants. Finally, the uniformity of the mixture contributes greatly to the properties of plastic, for usually the more homogeneous the mixture is, the better, or at least the more uniform, the properties will be.

METHODS OF COMPOUNDING

Strictly speaking, compounding involves the fusion of different materials into a homogeneous mass that is uniform in composition and structure. The actions in this type of mixing consist of smearing, folding, stretching, wiping, compressing, and shearing. However, component materials may be dry-blended either in preparation for compounding or for direct fabrication. With this in mind, let us consider the major types of plastics compounding equipment in current use.

Intensive Dry Mixers

Intensive dry mixers are used for dry-blending powdered resins, such as PVC with plasticizers and other additives or polyolefin flakes with pigment dispersions. A typical mixer consists essentially of a high-speed propeller-like impeller located at the bottom of a container. Heat generated during the blending cycle is continuously removed to stabilize the dry blend and improve flow characteristics. In one design the blend is cooled by impeller action. Heat is transferred from the constantly moving fluidized bed of material as it makes continuous and intimate contact with a cooling jacket wall. Other designs make use of cooling plate coils.

Impellers operate at the relatively high speed of 80 to 160 rpm. Moisture volatilized during the blending cycle is sucked into a vortex formed in the center of the mixer and is vented from there. High-speed chopper blades operating at about 3600 rpm can be used to break up agglomerates and lumps and thereby assure the rapid incorporation of fluids.

Large-capacity systems generally combine high-intensity mixing units with low-intensity cooling units. The cooling units usually exceed the mixing units in capacity to maximize operating efficiency. Mixing cycles are short, averaging 5 to 10 min.

Internal Intensive Batch Mixers

Internal batch mixers have been widely used in the production of vinyl film and sheeting since the introduction of these products. In fact, every large manufacturer of vinyl film and sheet uses internal mixers for dispersion and fluxing. High-speed operation results in overall cycles of 2 to 4 min. Internal mixers are also used for processing plastics such as vinyl, polyolefins, ABS, and styrene, along with thermosets such as melamines and ureas, because they can hold materials at a constant temperature.

Essentially, internal mixers consist of cylindrical chambers or shells within which materials to be mixed are deformed by rotating blades or rotors. In most cases a shell actually consists of two adjacent cylindrical shells with a rotor describing an arc concentric with each shell section. Because the rotor blade tips clear the cylindrical segment of the shell by a small amount, the blade motion causes the mixture to be sheared between blade tip and shell. The blades also interact between themselves to cause folding or "shuffling" of the mass. The shuffling is further accentuated by the helical arrangement of the blade along the axis of the rotor, imparting motion to the mass in the third, or axial, direction. Frequently, the blade is divided into two helices, pitched in opposite directions, in order to further the shuffling of components within the mixture. In some cases, the rotor blade arms are Z-shaped rather than elliptical. Still further variations are haben and fishtail blades. The shell and rotor are cored or otherwise provided with means for heating or cooling for the purpose of controlling the batch temperature.

The principle of internal batch mixing was first introduced in 1916 with the development of the Banbury mixer. Today's internal batch mixers, however, are far more rugged, durable, and efficient. The Banbury mixer consists of a completely enclosed mixing chamber in which two spiral-shaped rotors operate, a hopper to receive the material for mixing, and a door for discharging the batch. Revolving in opposite directions and at slightly different speeds, the two rotors keep the stock in constant circulation.

The ridge between the two cylindrical chamber sections helps force mixing; the acute convergence of the rotors and the chamber walls imparts shearing. This combination of intensive working produces a highly homogeneous mix. The batch is confined within the sphere of mixing action by an air-operated ram in the feeding neck. The ram also facilitates the flow between the blade tip and shell, where shearing is most extensive.

Internal mixers range in size from laboratory models with a capacity of about 2 lb of mixture to models of approximately 100 to 1500 lb. Comparable power ranges are from 10 to 100 hp.

The thermoplastic mixture discharged from the internal mixer is usually in the form of large shapeless lumps; it is frequently convenient to roll these lumps into sheets with a two-roll mill for eventual grinding or dicing, or to feed them directly into and extruder-pel letizer. A mill or extruder-pelletizer is then placed beneath the internal mixer – but, as a matter of fact, they may also perform additional mixing and dispersing.

More recently, mixers have been built with the capability of dewatering and devolatizing a variety of plastics. Slurries containing more than 70% moisture can be dewatered in 3 to 4 min. Most of the liquid is drained or pumped out of the machine; the remainder is vaporized by heat produced by mechanical action of the stock and contact with the mixer's heated working elements. Steam escapes through an exhaust system.

Color and formula changes can be made easily in internal mixers because mixing chambers are self-cleaning. Thus, a variety of stocks of different color or formulas can be handled in succession. Many internal mixers in operation today are used for repeated changes of color in the same machine – from white to green, to gray, to red, to brown, to black, all in the course of one day. The same adaptability applies to formula changes.

CONTINUOUS MIXER

The Farrel continuous mixer is one with rotors and mixing action similar to those of the Banbury. Raw material is automatically fed from hoppers into the first section of the rotor, which acts as a screw conveyor and propels the material to the mixing section where it undergoes intensive shear between rotor and chamber wall, kneading between the rotors, and a rolling action. An interchange of material between the two bores of the mixing section is an inherent feature of rotor design. The amount and quality of mixing can be controlled accurately by adjustment of speed, feed rates, and orifice opening. Since the feed screw is constantly starved and the mixing action is rotary, there is little thrust or extruder action involved. Stock discharge temperature is controlled by the rotor speed and the discharge orifice; production rates are controlled primarily by feed rates.

The list below shows the broad range of uses of continuous mixer:

- Polyolefin color masterbatching
- Polystyrene masterbatching
- Carbon black polyethylene masterbatching
- Polyethylene carbon black cutback
- Polyethylene homogenization
- Devolatilization of polyolefins
- Compounding of various ABS materials
- Compounding of plastic wood flour compounds
- Mixing various fibers into plastics

The continuous mixer can maintain optimum dispersion or mixing over a wide range of production rates. On most materials, a production-rate range of 5 to 1 or even 10 to 1 is possible. For example, a mixer designed to produce over 10,000 lb/hr PVC sheeting compound has proved capable of handling the same material, with equivalent quality, at rates of 1000 or 2000 lb/hr. Such operation is desirable in a calendering line where gauge changes in the sheet require different production rates.

In practically all mixing, discharge temperature is critical, either because it is a limiting factor, a measure of the amount of work done, or a goal. To maintain versatility, it is desirable for the discharge temperature to be independent of the production rates. The temperature of a continuous mixer can be adjusted within a range of 100°F at any given production rate and then maintained at the set temperature throughout the production run.

Generally, the mixer is fed – either with a preblend or with two, three, four, or more ingredients – on a continuous basis. But when the form of the materials or the proportions preclude accurate continuous-flow weighing and feeding, one or all of the ingredients can be fed in batch weighings. When these are fed at regular time intervals there is no loss in mixing or blend quality, and a continuous discharge is maintained. Sometimes minute quantities of an additive may require preblending with the primary ingredient before it is fed to the mixer. Sometimes a masterbatch may have to be prepared.

The continuous mixer is not completely self-purging, but one stock can be followed by another, without cleaning, with contamination of only a small amount of material. Still, before following with a different compound, it is generally desirable to clean the orifice unit, which is removable as a complete assembly.

Two-Roll Mills

As stated earlier, after being processed in a mixer, material may be dropped directly onto a mill for sheeting and dicing, or into an extruder or pelletizer. Arrangements vary. Where a mixer is located directly above the sheeting mill, the discharged stock drops into the bite of the rolls and can be taken off in a continuous strip. In other installations, where the sheeting mill is at a distance from the mixer, conveyors of various types are used to carry the mixed batch to the mill. High-production mixers require mechanical means of handling the large quantities of materials passing through them, both before and after mixing.

The two-roll mill consists of two opposite rotating parallel rollers placed close to one another with the roll axes lying in a horizontal plane, so that a relatively small space or nip between the cylindrical surfaces exists. Material reaching the nip is deformed by friction forces between itself and the rollers and made to flow through the nip in the direction of roll motion. Liquid plasticizers or finely divided solid ingredients are also placed in the nip and are incorporated into the resin through shearing action. Strip cutters are used to cut ribbons of stock from the roll for feeding to a cooling tank and dicer.

Usually, by adjusting roll temperature, the compound can be made to adhere to one of the rolls as a relatively thin sheet. The rolls are heated or cooled by a heating or cooling media introduced into their hollow cores. They are usually rotated at different speeds to facilitate the formation of a sheet or band on one of them. During mixing, the band is frequently cut and manually pulled loose from the roller. Generally, the gap between the rollers may be adjusted manually by means of hand-driven or motor-driven screws.

Roll mills vary greatly in size from very small laboratory machines with rollers of about 1 in. in diameter and driven by fractional-horsepower motors, to very large mills with rollers of nearly 3 ft in diameter and 7 or 8 ft in length and driven by motors over 100 hp.

Special-purpose roll mills are available with three, four, or five rolls, in which the material is caused to pass from one nip to the next, in succession. Likewise, rolls may be arranged in pairs within a single frame, each independently adjustable in clearance. Material is led through the various pairs in tandem in a cascade arrangement for continuous mixing.

Single-Screw Extruders

Although developed to form thermoplastics, the single-screw extruder also functions as a mixing device, because it subjects materials to laminar-flow deformation. Thus, although the extruder is not primarily used as a mixer, it is frequently used to add ingredients to a resin during melt extrusion, thereby taking advantage of its inherent mixing action. It is quite common for colorants to be added to extruded products in this way. Both the colorant and resin are metered into a feed hopper, and a suitably colored extrudate is frequently obtained.

Often, reclaimed scrap stocks of different colors are mixed satisfactorily in the extruder provided that the materials are suitably metered in the feed hopper. Again, a masterbatch or premixed concentrate of colored thermoplastics may be fed to the extruder simultaneously with the resin to be colored. Here, the extruder becomes an auxiliary mixer.

The amount of mixing a given volume of resin receives may be expressed in terms of the total amount of shear to which the resin is exposed. It is the product of shear rate and the extruder residence time, and a measure of the relative displacement of one particle with respect to its neighbor.

The total amount of shear in an operation can be increased in one of several ways: by diminishing the channel depth of the screw, by letting the screw helix angle approach 0° or 90°, or by increasing the amount of pressure flow and leakage flow through increased die restriction.

The total amount of shear to which a particular element of liquid resin is exposed depends upon its initial position in the screw channel, since flow path, and thus both residence time and shear rate, vary according to the initial position of the material in the cross section of the screw channel. Material near the center of the channel receives less mixing than material near the screw surface or the barrel wall. This is because of the shorter residence time of the material near the center of the channel. It is also important to note that no motion of material backward along the screw axis ever takes place. Hence, single-screw extruders provide no "turnover" and little bulk mixing in the axial direction.

Many types of special mixing heads, or torpedoes, have been designed as attachments to the end of extruder screws. All have the primary purpose of further increasing the amount of shear to which the extrudate is exposed and of crossblending the somewhat nonuniformly mixed material leaving the screw channel.

There are different kinds of conventional single-screw extruders for different operations. In the area of compounding, they can be conveniently grouped under the following headings: mixer-fed extruders, hot-melt extruders, and color-feed extruders. Available head arrangements for each type include strip, strand, slug, cone, pelletizing, and underwater pelletizing.

Mixer-fed extruders are usually located directly under a batch or continuous mixer and fed by gravity.

Operation is automatic; a variable-speed drive is used to synchronize the extruder speed with the production rate of the mixer in order to keep the screw always full of stock and the discharge uniform and unbroken.

Extruder cylinders are generally jacketed for steam or liquid heating, but they may be electrically heated, as required. Extruder size is dependent on the production rate of the continuous mixer or the size and mixing cycle duration of the batch mixer. Drive requirements are determined to a great extent by the compound to be handled.

Hot-melt extruders take the resin from materials suppliers' reactors or receivers and deliver it in flat-strip or pelletized form. A great many hot-melt extruders have been built in recent years, most of them for handling polymers within the range of 0.2 to 50 melt index.

Extruder size is determined by the hourly production rate desired. Horsepower requirements, throughput rate, and screw design are dependent on material characteristics.

Cold-feed extruders are available for working high-density and low-density polyethylene, plasticized PVC, and other compounds. The machines are generally cube- or pellet-fed and deliver extrudate to pelletizing heads. Most have completely jacketed construction, designed for use with pressurized water-circulating systems. They are started up with circulating hot water and then, as material flow is established, water temperature is gradually reduced to obtain the best operating conditions for the resin being handled.

Compounder-Extruder

One modification of the single screw is the Transfermix (Sterling Extruder), a single-step "enforced order" variable-intensity compounder and extruder operating over a stepless gradient from extensive to intensive mixing, with a running cutoff on the intensive-mixing end at any desired shear rate between 30 and 3000 reciprocal sec.

The unusual design incorporates a rotor, or single rotating screw, inside a series of stators, or stationary screws, which replace the smooth barrel of a conventional extruder. The stator and rotor grooves are opposite-handed and alternatively increase or decrease in depth. The grooves in both rotor and stator are shaped to help the material vortexing.

The cross-sectional diagram of the rotor and stator shows six extensive mixing stages, a vent section for removing volatiles and entrapped air, and a wide-landed, conical, high-intensity, variable-shear section. The material starts to vortex to the grooves of the rotor as a solid as soon as it reaches the barrel. Since it is immediately transferred to the stators where it makes contact with the heat-transfer walls of the stator grooves, it starts to melt and immediately is mixed with the bulk of the solid material, melting further resin.

This transfer of material from rotor to stator, and vice versa, can be broken down into different actions, namely, vortexing and changing the flow or direction. In vortexing, the vortex of material in the giver screw diminishes and increases because, in the taker screw, the outside layer of the vortex of the giver screw unpeels and then becomes the ouside of the vortex in the taker screw. In this way, all of the material comes in contact with the heat-transfer walls of both rotor and stator, once per stage. (A *stage* is defined as one transfer from rotor to stator and back again.)

Besides the vortexing in the grooves, the material in the rotor moves forward, more or less in a straight line or gentle helix. But as material is transferred into the stator, it must also follow the exact helical path of the stator. Thus, there is a change in direction or flow with every pass from rotor to stator and back.

It is this change of direction of flow that is responsible for the pumping or extruding capability of the machine regardless of the nature of the material. If the material tends to turn with the rotor, it is nevertheless pumped forward by grooves of the stator. If it does not turn with the rotor, it is propelled forward like a nut on a bolt.

The Transfermix is used in the following applications:

- Color concentrate and carbon black masterbatch
- Direct mixing of low concentration of pigment in resin
- Filling of resin (talc, asbestos, glass, etc.)
- Resin/resin compounding
- Resin/additive compounding for special properties
- Low-shear compounding of cross-linkable polyethylene and precharged foamable pellets.

Twin-Screw Extruder

In twin-screw extruders, two screws are arranged side by side. One design incorporates corotating screws that are intermeshing and self-wiping. Since the screws rotate in the same direction, material moves helically along the inside barrel wall in a figure-8 path from the feed section to the discharge point.

The geometry of the screw components is such that the root of one screw is constantly wiped by the flight top of the second screw with a uniformly small clearance between them at every point. Thus dead spots are eliminated, the residence time of each melt particle is uniform and purging times are shortened. Because of the positive conveyance of material, all types of material – molten polymers, pastes, flakes, powders of low bulk density, and fibrous materials – are efficiently conveyed irrespective of friction coefficients. With positive conveying, operations with partially filled screw flights are possible, while degassing is going on in some zones or additional components are introduced into others.

Screw configurations are used to vary conveying efficiency, throughput rate, the degree of filling, and pressure build-up. Screws with reversed flights are used to generate localized high pressure. Hydraulically operated dynamic values between barrel sections can be used to allow pressure variations during operation. Staggered-stepped screws provide intensive transversal mixing by shear forces of varying intensity. Various mixing and kneading effects are obtained by a suitable screw design.

Residence time is determined by barrel length, screw lead, screw speed, and throughput rate. Residence time may vary between 20 sec and 10 min., depending on the process and operating conditions. Residence-time distribution may be influenced by changing the screw geometry, by which residence-time distribution can be widened to handle considerable longitudinal mixing.

Selective removal of heat from the melt by cooled screws and barrel walls results in increased melt viscosity and, consequently, in higher shear rates. The need for external heating can be minimized by appropriate screw geometry. Depending on product and process, the specific input energy may be varied between 0.05 and 1.2 kWh/kg.

The material in the barrel is worked into thin layers by relatively narrow screw flights. The design increases processing intensity, particulary at the barrel crest, where the material transfers from one screw to the other, and the layers are continuously mixed and inverted.

Volatiles are removed through vent ports at different locations along the barrel length. Fluids or melts of high or low viscosity or free-flowing solids may be introduced into the barrel at various points by means of suitable metering devices.

Twin-screw extruders with counterrotating screws are also available and, despite similarities, there are some significant differences in the way they handle the melt. What are some of them?

With counterrotating screws, screw flights carry material by friction in such direction that all of it is forced toward the point where the two screws meet, there forming a "bank" of material similar to that of a two-roll mill, although some material slips through the gap between the two screws. Like the two-roll mill, the theory is to feed material through the nip from the bank on top of the nip gradually and statistically, in order that each particle is processed equally over a period of time.

Counterrotating twin screws can put the material passing through the nip under an extremely high degree of shear, in the manner of a two-roll mill. Varying the clearance between the screws varies the portion of material fed between the screws against the portion of material accumulated in the bank and simply moving down the barrel. Material passing through the nip can be subjected to great shear forces by narrowing the clearance; also, the portion of material in the bank is greater.

Statistically, however, only one portion of the material is subjected to the high shear condition of the nip and another portion only to the low shear condition of the bank. Counterrotating twin screws are not totally self-cleaning.

In corotating twin screws, one screw transports the material around up to the point of intermeshing, where, because of the existence of two opposing and equal-velocity gradients, a great majority of the material is transferred from one screw to the other along the entire barrel length in the figure-8 path mentioned previously. Because the figure-8 path is relatively long, the chances of controlling melt temperature are much better.

The amount of shear energy developed at the point of deflection can be regulated within very wide limits by choosing the depths of the screw flights.

Self-cleaning screws, aside from the obvious advantage of facilitating color change, provide control over residence-time distribution. Control of residence time is also of great importance for heat-sensitive resins and pigments, or for operating at higher processing temperatures. A short and uniform residence time is essential to minimize heat and shear history, and, thereby, maximum quality.

On-the-Machine Compounding

Recently, equipment and materials have been made available which make it possible to mix the colorants and additives with plastics directly on the processing machine in a continuous operation. These techniques are particularly suitable for extrusion of colored plastics and fall into four broad categories: 1) hopper-mounted color mixers; 2) liquid color-metering processes; 3) melt proportioning; and 4) static mixers.

Hopper-Mounted Color Loader/Mixer

Hopper-mounted coloring loaders combine virgin materials, regrind, one or two colorants or additives (slip agents, inhibitors, etc.), mix them by tumbling or by gravity, and drop the mixture into the process hopper. Some units allow use of dry powered colorants or color concentrates without major equipment alterations. The coloring units are self-loading and mount directly over processing so there is no need to manually handle virgin, regrind materials, or additives, and risk contamination and waste. Dry pigments or color concentrates are placed in a canister in a separate color room and the filled canister is carried to each machine as needed.

While the coloring loaders minimize labor and time costs during color changes, the machine hopper has always presented a major downtime factor because it has to be wiped clean before the next color is used. To get at the machine hopper it is usually necessary to move the loader out of the way, which of course slows the process further.

A new clear plastics cone, held in a steel frame, is available for mounting either a regular hopper loader or a coloring loader. The cone is of low cost so it can either be thrown away or saved for the color the next time it is run.

Color concentrates are usually conveyed automatically into a color storage bin on the coloring loader. The colorant-metering screws meter in the desired percentage of color to each fixed volume batch of natural virgin material. Freeze-dried colorants, which are conglomerations of powdered dry colors in a cleaner and more easily handled particle form, can be successfully metered through standard coloring equipment and, in fact, are preferred in some cases to powdered dry colors because of their cleanliness.

Metering accuracy is important in colorant loaders, as is their ability to handle a wide range of material types and combinations. It is usually advisable to pretest materials through these units before deciding if they will meet processing requirements. Consideration of regrind and thorough incorporation of colorants with both virgin and regrind materials are important to successful on-the-matching coloring, regardless of the type of colorant.

Liquid Color Mixing

Two basic methods of using liquid colorants are currently offered. One, which is an interchangeable fixture for a coloring loader, meters a precise amount of the colorant into a precalibrated charge of virgin and/or regrind in the unit's mixing drum. All materials, additives and colorants are thoroughly mixed together before they discharge into the hopper. Another method of liquid coloring includes a pump which feeds a controlled amount of colorant directly into the screw of the injection machine as virgin materials are flowing into the throat. The latter method is more popular.

The total-system approach of liquid-colorant addition affords precision at-the-throat metering for operations such as injection molding and extrusion. The liquid additives are metered directly from one or more containers into the throat of the machine by a precision metering device whose operation is mechanically or electronically integrated with that of the rotating screw. A variety of metering pumps, designed for handling the usually high viscosity of the liquid additives, is on the market. The metering pumps are normally tied into the screw-return cycle for injection molding processes to permit uniform color addition over the entire charging cycle. For extrusion, the pumps are tied into the screw drive mechanism to compensate for possible speed fluctuations.

There are several types of metering-pump designs available, with alternative features. The basic designs include the positive displacement piston type, the diaphragm-head type, the piston-actuated diaphragm type, the high pressure screw auger type, the peristaltic type, the gear pump type, and others. Choosing the best metering pump metering unit for a given application can be made easier when the control parameters for the process are identified; part size or output rate, proximity of feed source to equipment, and available power supply to operate the metering device.

The utilization of liquid-color mixing for continuous extrusion processes and in plastisols seems to be straight-forward. In fact, this is the area where liquid-color mixing has now become firmly established. The use of liquid colors in injection molding is relatively new and is still in a developmental stage. A few remarks are therefore warranted in this connection.

Correct placement of the liquid-feed tube is critical in reciprocating-screw injection-molding machines, in order to obtain proper at-the-screw metering and mixing of the liquid with the resin being fed from the machine hopper. The general rule is to place the tube in the center of the vortex of resin about 2 to 3 in. above the screw. In this manner, the resin carries the liquid without bridging, and the liquid is certain to wet the resin before it touches the screw. In some cases, if the tubing is placed too close to the screw or the side of the throat, the liquid can aggregate on the inner screw flights causing channeling and distribution problems in the finished part. One way to assure proper tube placement is to use a space plate between the hopper and the throat with a brazed section of copper tubing bent to terminate at the correct location. A disposable length of plastic tubing is then inserted into the copper tubing until it hits the stop, thus assuring correct tubing placement and rapid, contamination-free color changeover.

Another important consideration in injection molding is the accurate phasing of liquid addition during the resin-charging cycle, since resin flow is intermittant. The metering device must start discharging liquid the instant the screw begins to run, and stop exactly when the screw stops. Such coordination is obtained by the use of a relay control that ties the metering pump operation to the screw-return cycle. Some compensation has also to be made for after-dribble, as is the case when long lengths of flexible tubing are used for metering from the pump to the throat. This after-dribble is caused by the swelling of the tubing, and can be minimized by using a more rigid tubing. Precalibration of the metering pump is recommended so that the correct let-down ratio can be set prior to start-up. However, with a little familiarization, a particular unit can be preset based on the known displacement vs. the known specific gravity of the liquid colorant. Once a machine is set up and operational, the level of colorant can be quickly adjusted to compensate for opacity, or to compensate for various regrind levels.

For extrusion processes, placement of the liquid-discharge tube has somewhat the same structures as indicated for injection molding. The metering device can be tied in to the screw speed by a dc drive motor and circuit, or by a mechanical drive while on the screw shaft. To optimize color distribution, liquid feed should be as continuous as possible. In the case of pulsating pump units, it is recommended that at least one pulse occurs for each screw revolution.

Screw slippage has been considered by some processors as the greatest drawback of using liquid colorants. However, most of these problems were exaggerated by improper introduction of the liquids, causing the feed section of the screw to be immersed in the liquid. (In some cases the reverse has been experienced.) The liquid colorant dispersions can alter the rheological properties of the melt to the extent that 10 to 20% increases in the output rate may occur. The introduction of liquid colorants in both cases is usually accompanied by a decrease in screw power input at a given set of conditions. This drop in power consumption results from the liquid-colorant dispersion's lubricating effect during the melt state.

Another method of liquid-colorant introduction in extrusion processes has received considerable attention recently. This method is a high-pressure injection of liquid colorants directly into the plasticated material just beyond the feed section of the screw. Thus, The feeding capacity of the feed section is not affected, while the colorant dispersion acts as an internal lubricant, with an increase in output rate. The correct area of introduction for a given machine is forward of the feed zone, so as to not affect efficiency, but close enough to the feed throat to utilize as much of the compounding capability of the screw configuration as possible. The liquid is usually introduced through a standard thermocouple-probe port, sometimes used as a grease-seal pressure-gauge port. An adapter and check valve assembly are used to prevent back flow of the melt plastic into the pump lines. Attached to the assembly is a high-pressure feed pump capable of handling the normal high viscosities of the liquid colorants at discharge pressure exceeding the 500 to 1300 psi, normally occurring in the area of the injection port. With this process it is possible to accommodate the relatively high con centrations of additive liquids required for the fabrication of foamed profiles, blown film, and thin gauge sheet.

Melt-Proportioning

Hopper-mounted color loading and liquid color-metering techniques are quite adequate for extrusion of colored plastics for high throughput operations and for noncritical products. For more critical end uses, such as the extrusion of fibers, film, code ribbons, wire coating, etc., a high degree of color uniformity is desired, and this can be achieved by continuous color-mixing operations through the use of a melt-proportioning technique.

In melt proportioning, the color concentrate is melted in a slave extruder which feeds it into the main extruder through a gear-metering pump. A precise dosage of color is therefore possible. The point of entry of the molten color concentrate into the main extruder is generally in the transition region of the screw. It is essential in melt proportioning that free-flowing color concentrates, and not the pure pigments in color dispersions, are used. A 10:1 letdown is possible but 7:1 or lower is advisable. The melt-proportioning technique is the ultimate in achieving color uniformity, but the hardware is rather expensive.

BIBLIOGRAPHY

1. Mohr, W. D. "Mixing and Dispersing." in E. C. Bernhardt (ed), *Processing of Thermoplastic Materials.* New York: Reinhardt Publishing Co., pp. 117–152 (1959).
2. Bikales, N. M. "Compounding." *Encyclopedia of Polymer Science and Technology* **4**, pp. 118–129, New York: Interscience (1966).
3. Dohertz, R. B. "In-Plant Coloring of Plastics: Procedure, Pitfalls and Payouts." SPE Regional Technical Conference, Cherry Hill, New Jersey (Sept. 1973).
4. Hackney, H. U. "Liquid Colorants for Thermoplastics." SPE Regional Technical Conference, Cherry Hill, New Jersey (Sept. 1973).
5. "Processor's Guide to Mixing, Blending and Compounding Equipment, A State-of-the-Art Report." *Plast. Technol.* (July 1967).
6. Adams, R. L. "High Volume Production of Pigmented Thermoplastics." SPE Annual Technical Conference, New York (May 1970).
7. Murray, B. G. "Automatic In-Plant Coloring, The State-of-the-Art." SPE Annual Technical Conference, New York (May 1970).
8. Adams, R. L. "Producing Pigmented Thermoplastics in Volume." *SPE J.* **27**, pp. 54–57 (Apr. 1971).
9. Schneider, M. Z. "Color Concentrates for Thermoplastics – A Different View." SPE Annual Technical Conference, New York (1970).

10. Robbins, G. H. "The Utilization of Color Concentrates." SPE Annual Technical Conference **XXX**. Chicago (1972).
11. Adams, R. "Pigmentation of Thermoplastics Using Twin-Screw Compounders." SPE Regional Technical Conference, Philadelphia (1972).
12. "Does In-Plant Color Blending Pay?" *Plast. World*, pp 44–46 (July 1972).
13. Paryzek, J. C. "Concentrates for Blow and Cast Polyethylene Films." *Tappi* **56** (7), 126–128 (July 1973).
14. Schoengood, A. A. "Compounding." *SPE J.* (Feb. 1973).
15. Hackney, H. U. "Reconsider Liquid Coloring for Plastics." *Plast. Eng.*, pp. 61–63 (Mar. 1974).
16. Kennedy, R. J. "Internal Pigmentation of Low Shrink Polyester Molding Compositions with Flushed Pigments." SPE Annual Technical Conference (1975).
17. Forger, G. R. "Filling the Additive Hopper – A 1977 Update." *Plast. Eng.*, p. 20 (May 1977).
18. Snow, J. P. "Using Liquid Color to Best Advantage." *Plast. Eng.*, p. 17 (June 1977).
19. Stone, H. "Process Coloring of Thermoset Resins." SPE Annual Technical Conference (1977).

9
Surface Coloration of Plastics

INTRODUCTION

Apart from mass pigmentation, as discussed so far, there are three other methods available for the coloration of plastics (painting, printing, and dyeing). The volume of plastics colored through these methods is not as great as that colored through mass pigmentation; nevertheless, these are important methods of decorative coloring where performance requirements allow and when cost structure is acceptable.

PAINTING

It is estimated that painted plastics account for 5% of the plastic parts produced in the United States, with major areas of concentration in automotive, furniture, marine, appliance, building and construction, and business machine applications. The benefits to be obtained by painting plastics are both decorative and functional. Among the decorative benefits are: a greater range of colors; the possibility of altering a product's surface appearance; a less plastic image; reduction of gloss; and decorative effects such as polychrome, woodgrain or pearl finishes. In addition, painting is the only method by which light-colored finishes can be obtained from dark resins such as phenolics. The functional advantages of painting plastics include:

covering defects such as flow lines; improvement of weather and wear resistance; and additional resistance to chemicals, water, solvents, etc. Some plastics, like ABS, are susceptible to ultraviolet degradation. Painting the ABS part overcomes UV degradation and upgrades its appearance. Another instance of dual benefits is the painting of open or closed cellular urethane foams. Here, painting imparts color stability and weather resistance. Painting of self-skinning foams overcomes their tendency to yellow. Painting also permits the use of mixed residues of regenerated plastics of various colors.

One of the most critical requirements of painting is good adhesion with no embrittlement. Adhesion is generally determined by the standard crosshatching – good adhesion is indicated by no removal of paint. Embrittlement is generally observed on cutting the paint film before taping. In evaluating new products it is advisable to recheck adhesion after a period of a few days.

Unlike finishes for other surfaces, paints for plastics require careful selection of solvents, a critical feature in their formulation. Restriction in the solvent, in turn, sets a limitation on the resin-coating system that can be adopted for any specific plastic. Thermoset plastics are resistant to most solvents. Thermoplastics, however, are generally sensitive to solvent attack to varying degrees. A common defect is solvent etching of thermoplastics, which renders their appearance generally unacceptable. In more extreme cases the solvents can distort or even disintegrate the plastic. Attempts to obtain adhesion by solvent etching tend to give inconsistent results including dull spots in low-density areas. A solvent system should, however, provide good wetting of the plastic. A common defect on painted plastics is crazing or cracking of the molding. This appears to be related to the effect of the solvent on releasing mold strains. Correct solvent formulation, annealing and possibly molding conditions, will correct this defect. In general, strain-free molding is required with thermoplastics to avoid distortion during baking, particularly at higher temperatures.

For satisfactory painting, no mold-release agents should be used. Such contamination can lead to problems of surface defects, such as poor adhesion. In painting urethane foams, the residual mold release has a particularly adverse affect on adhesion and must be removed. For removal of silicones, strong solvents such as trichloroethylene or perchloroethylene have been suggested. Some plastics contain internal lubricants that are soluble in some of the solvents used in coatings; as the lubricants increase, finishing problems can be aggravated.

Most plastics have a tendency to become electrostatically charged, leading to dust attraction, marring of appearance, and, in more extreme cases, disturbance of flow characteristics of the lacquer. The electrostatic charge can be removed or at least reduced by washing with a suitable solvent destaticizer. The parts are simultaneously cleaned, and the solvent must be allowed to evaporate rather than be removed by rubbing the parts, which would recharge them. Alternatively, ionizing air guns can be used to remove dust and other contaminants and to prevent reattraction. Both procedures are effective for only a limited period and are therefore best employed just prior to lacquering.

Certain difficult plastics, such as polyolefins, require drastic pretreatment to attain satisfactory adhesion of the coating. For example, polypropylene and polyethylene are pretreated by flame, acid, or vapor etch. Alternatively, they are prepainted with a dilute solution of chlorinated polyolefins, such as those marketed by Eastman. In flame treatment, a hot oxidizing flame (1100° to 2800°C) is impinged for a minimum time on the surface to be draw rated. On irregularly shaped articles there may be difficulty in achieving effective treatment over the entire part. In acid treatment, the parts are immersed for 3 to 5 min. in an acid solution, then thoroughly washed and dried before coating. Shorter times are used for similarly pretreating acetal polymers, but care is required to avoid crazing. For vapor or solvent treatment the components are immersed for 15 to 30 sec in a hot solvent, preferably perchloroethylene, trichloroethylene, or toluene. Unlike the acid and flame processes which change the chemical nature of the surface of the plastic, vapor etching appears to lead to swelling and roughening of the surface. Distortion of the component can occur and it must be coated within a certain time; otherwise, the beneficial effect of the treatment is lost.

To apply various adjoining decorative effects such as on badges, horn buttons, appliance escutcheons, name plates, trade marks, etc., very accurate masking is essential. The best results are obtained with electroformed masks, one for each decorative step. In designing the molding, the need for subsequent masking should be borne in mind by allowing for stops to hold the masks.

Spray application is the most versatile and frequently used procedure for applying coatings to plastics. Manual spraying with suction or pressure feeding and manual handling are still widely used, particularly for smaller runs. But even for smaller production semiautomation is possible (the spray gun or masks being operated by foot pedals). Automatic spraying is widely used for high-speed production of large runs.

An interesting product introduced for use with polyurethane foams is a polyurethane barrier-release coating. It is designed for spray application to most types of polyurethane molds to protect them from attack by the foam. Drying takes only a minute or two. The polyurethane foam is then foamed into the mold and after curing is readily removed. The coating has excellent release properties from the mold and adheres to the polyurethane. It is suitable as a finishing coat or may be used as a ground coat for any further decoration, such as wood graining.

Wiping paints are used for decorating indentations and debossed letters and designs. Their use for these purposes is a simpler, more economical procedure than masking. *Silk screening* of plastics is accomplished in the same manner as it is for decorating metal, glass, or other substrates. Normally, flat surfaces are silk screened, but curved and cylindrical molding may be, also.

Highly decorative effects simulating wood graining can readily be applied to most plastics. A ground coat of color is applied directly to the plastic (pretreated, if required) by roller coating or spray application. The ground coat is generally quick drying (10 to 20 min.) and may be force dried for a few minutes on a production line. A graining ink of contrasting color is then applied from a graining plate. A roller, wetted with the selected graining ink, is rollered over the ground coat. The graining coat is followed by a coat of lacquer, either gloss, eggshell, or flat. The final coat of lacquer, which gives a uniform sheen, imparts a realistic wood finish.

Paints for plastics are air-dried or baked at elevated temperatures, depending upon the specific formulation. Typical temperatures at which some plastics may be baked without mold distortion are: polystyrene, 140 to 150°F (60 to 65°C); acrylic, 150 to 175°F (65 to 80°C); ABS, 175 to 195°F (80 to 90°C); thermosets such as phenolics and urea, 230 to 300°F (100 to 150°C); polycarbonate, polypropylene, polysulfone, and phenylene oxide-based resins, 265 to 285°F (130 to 140°C). Ovens that are solvent laden, with uneven temperature areas, do not give good results and can cause defects. Both batch and conveyed ovens should be well ventilated.

In view of the recent enactment of solvent emission regulations under the Clean Air Act, increasing interest is being shown by the plastics industry in solvent-free coatings. However, despite a furious spurt of R and D activity by the coating industry, solvent-free coating systems suitable for plastics are just starting to reach the market. In all cases, the alternatives to solvents – water-based coatings, radiation-cure coatings and powder coatings – still present some problems when applied to plastics. A cure-all coating system that the processor could simply plug into his existing coating line has not yet arrived.

The coating technology that has generated the greatest anticipation in the plastics industry has been water-based coatings. The principle reason is that water-based coatings could be directly substituted for solvent-type materials and used on processors' existing coating lines. But while coating manufacturers have had reasonable success in developing water-based systems for the big coating markets (metal and wood), developments have been slower in plastics. What has slowed progress is the fact that the polymers on which water-borne coatings are based generally cure at fairly high temperatures (about 350°F). This limits the number of plastics on which they can be used. The problem, of course, is getting the polymer in the coating (usually either polyesters or acrylics) to polymerize at low temperatures.

Work is still in the development stage; PPG, M & T, Rohm and Haas, Du Pont, Union Carbide and others have active programs. Some products are reaching the market or, at least, are being made available to coating formulators.

A totally new technology for coatings, based on a long-known phenomenon – that radiation will cause monomers to polymerize – is getting as much, if not more, attention as water-based coatings. In fact, many experts consider water-based coatings to be only an interim measure, to be used while processors get ready to accept a new coating technology and are able to afford the necessary hardware. Their reasoning is based on the fact that the amount of energy required to drive the water out of water-based coatings is much greater than that required for organic solvents.

At the present time, technology exists for curing coating with two types of radiation: UV or EB (electron beam). Presently most of the activity is concentrated in UV curing, which has the principal advantage of lower equipment costs compared to EB. However, EB operating costs are by far the lowest of any coating-cure system currently available. As a result, there are an estimated 20 UV curing lines for plastics in place compared to less than 10 EB lines. However, as far as replacing paint on plastics parts is concerned, UV curing has limited applicability. Because of the low energy levels of UV, it is generally suited for clear coatings up to one mil in thickness and for pigmented coatings in the range of 0.5 mil. For this reason UV is mostly being used to cure inks in printing applications on plastic films or sheet. Also, since all surfaces to be cured must be "seen" by UV radiation, it is limited to curing flat, or nearly flat, surfaces.

Electron beam, on the other hand, is capable of curing thick, heavily pigmented, paint-like coatings. The energy levels of the electron bombardment are high enough to penetrate thick coatings and, because the electrons set up inelastic collisions that propagate through the coating, it is more suitable to coating irregular surfaces. That EB curing is a viable alternative to solvent-based paints for plastics parts is illustrated by the success of Ford's Electrocure process. The EB line, which was installed at the company's Saline, Michigan, plant in 1970, is curing coatings on molded PP and ABS dash panels, knobs, trim parts, etc.

The technology behind coatings for UV and EB curing is essentially the same. The coatings are based on a variety of polymerizable monomers. Polyesters and acrylics are the most common, but urethanes, alkyds, epoxies, and silicones also are involved to a lesser extent. The principal differences between the formulated coatings is that those for UV curing must contain a photosensitive initiator that will catalize the polymerization reaction. These initiators add an estimated 15% to the cost of UV coatings compared to EB coatings, which require no initiators.

Radiation-curable coatings are under development by most coating and coating-chemical manufacturers. While much of the current R and D for UV is aimed at inks for web-printing applications, pigmented industrial coatings for EB are on the way.

The direction that development in EB coatings is taking is toward coatings that will bond chemically to the plastic substrate. For example, Fuller-O'Brien has developed an EB coating for ABS in automotive exterior applications. The coating is said to have withstood 2400 hr of salt-spray treatment without failing and has withstood rupturing of test plaques without delamination.

Aside from high equipment cost, the main factor slowing the adoption of EB for curing coatings on plastic parts is the fact that existing coatings are difficult to handle in spray equipment. Also, there are possible toxicity questions surrounding some of the monomers that are in use. While Ford is using an electrostatic spray system to apply its coatings for Electrocure, precautions for protecting operators have been taken. As a result, EB (and UV) coatings are generally applied by flow coating or dip coating, both of which are more suited to flat surfaces.

Increasing interest is evident in the electrostatic spraying of plastic moldings, particularly open moldings. Since plastic components are nonconductors, prior application of a special solution renders the molding conductive and permits electrostatic spraying. The temperatures required to fuse the powdered polymers used in electrostatic powder coating will keep that painting technique out of many plastic applications. However, development is underway to produce powders that will cure at lower temperatures, and powder coating is finding a significant number of plastics painting applications. Experts predict expanding use of powders in plastics applications for several reasons: powder coatings make more efficient use of coatings because electrostatic attraction eliminates overspray; no solvents are used and energy requirements are reduced.

Curing temperatures are coming down. Most of the early powders on the market required temperatures approaching 400°F to fuse, but companies are now reporting fusion temperatures in the 300°F range. The M & T Chemical is reportedly working on crosslinking agents that will bring cure temperatures down to nearly 200°F.

With the temperatures that are achievable at present, powder applications are limited pretty much to thermosets and some engineering plastics, but a number of potential applications exist in those areas. Detroit automakers and truck companies are said to be looking at powder coating for RP and some ABS parts. International Harvester spent a year on evaluation of a RP truck body that was experimentally coated with an acrylic powder supplied by Grow Chemical.

While RP parts probably represent the most immediate potential powder coating on plastics (and numerous development projects are underway), both Polymer Corp. and Farboil are offering pigmented epoxy powders for coating phenolic parts for appliances.

PRINTING

The processes commercially available for printing on plastics are many and varied. It is important that the properties of the part (dimensions, configuration, and type of material), its end use, and required output be carefully considered when choosing the printing process. The processes vary in the shapes that they can print, the types of inks or pigments used, and the degree to which they can be automated. Conversely, the part design should be carefully examined to see that it does not exclude the use of the most economical printing method.

Screen printing involves use of a metal wire or a nylon mesh stencil prepared by a photographic and chemical process. A squeegee or applicator forces paint or pigment through open portions of the screen onto the surface of the container. The thickness of the ink is such that screen printing can usually be identified by running a finger over the imprint to detect the raised effect. By contrast, hot stamping will be recessed and offset decoration will seem perfectly flat.

A wide range of opaque colors and high-quality production on plastics is offered by screen printing. It is an excellent system for plastic bottles because of the light pressures required. Internal support is easily provided, even for thin-walled containers, by the use of low-pressure compressed air.

Special pigments give excellent adhesion on a wide variety of plastics by creating a chemical bond with the container surface. After screening, the bottles are dried in conveyor-type ovens, which facilitate rapid handling.

Major screening improvements include fast-drying pigments that do not require post-decorating oven drying. The decorated bottles are ready for shipment as they leave the screening machine. In cases where some drying and curing are necessary the screened containers are conveyed past a small hot-air blower, adequate to dry and cure the pigment prior to packing. These fast-drying pigments eliminate the need for space-consuming drying ovens, an inherent part of screening in the past.

Screen printing can be adapted to either flat or cylindrical items, as well as oval and cone shapes. For flat objects the screen is held stationary over the part while the squeegee moves laterally to push the ink through the pores. For cylindrical objects the screen is moved laterally between the squeegee and the item, causing the item to rotate as the ink is applied.

The Tampo-Print system represents a significant plastics printing breakthrough. This machine (developed in Europe and now operational in the United States) allows coverage on surfaces and contours that previously were impossible to decorate. The process involves the use of a debossed metal plate and a soft conical-shaped transfer pad. The printing is acid-etched into the metal plate. This etched area is filled with pigment by moving a brush that continually carries the pigment from a tray-type reservoir.

A doctor blade cleans the surface of the plate to assure that no pigment remains in the nondecorating areas. A soft silicone rubber transfer pad comes in contact with the plate and picks up the pigment from the etched area. The transfer pad then comes into contact with the object to be printed and transfers the design to the decorating surface. All of the pigment is transferred to the object; the pad is almost completely clean after transfer.

The system does not require the heavy application pressures common to other decorating systems. This permits printing of highly sensitive and thin-walled products. Further, since there are no heavy application pressures involved, the need for expensive and complicated supporting jigs and fixtures is also eliminated.

Manual production rates range up to 2200 parts/hr, and automatic systems can produce up to 20,000 parts/hr, depending on the nature of the product. The unit has been used to put designs and imprints in such places as the insides of ashtrays, sides and bottoms of thin-walled bottles, on convex and concave containers, foam drinking cups, and dolls' eyes. And it can be used to decorate molded parts with widely varying external dimensions.

When being decorated with screen-printing pigments the objects must be dried immediately after decorating. Some types of paint can be air dried, others require longer curing in conveyorized ovens. Some fast-drying pigments permit safe handling of the parts immediately after decorating.

Halftone effects and multicolored halftone designs are also possible with this new system. This is accomplished with halftone plates and the use of several ink reservoirs. The transfer pad applies each color separately. This is done without drying between color applications and in rapid sequence. The pad deposits its entire load of ink on the item and is virtually clean of ink after transfer. Since the pad will not pick up ink from a decorated part, several layers of pigment can be applied to the same area without smearing or distorting the pigment already present.

In offset printing, a drum bearing a printing plate with raised lettering comes in contact with an inked roller. The ink picked up by the raised impressions is then transferred to the flat surface – generally of rubber – of the printing drum. As the item is rolled across the surface of the printing drum it picks up the printing. By arranging several inking plates around the printing drum, each using a different color, multicolored designs can be applied economically at high volumes. But adhesion and abrasion resistance of offset inks on most plastics are inferior to both hot stamping and silk screening, and the quality and richness of colors are not so true or opaque.

Several methods are available for transferring multicolored printing or designs from a roll-fed web to a plastic part by applying heat and pressure.

The equipment used in heat-transfer processes is either similar to or the same as that used in hot-stamping metallic foils. These systems involve the use of a release-coated carrier (paper or plastic film) upon which a design is printed. The design is later transferred from the carrier to a container, using either heat, pressure, or both.

The main advantage of heat transfer is that it allows the use of multicolored designs at relatively low cost. High-speed processes are used to print the decoration on the carrier and all colors are later transferred to the container simultaneously. Several transfer systems are available for plastic containers; these vary by the type of carrier used and the equipment required to effect the transfer.

Therimage transfers, a development of Dennison Manufacturing Co. (available since 1957), were designed primarily for decorating blow-molded bottles at speeds up to 120 per min. The speed and efficiency of this system gives it widespread application for many high-volume containers, including tubes, vials, and jars.

Basically, Therimage involves printing a design on a patented release-coated paper. The use of five-color rotogravure printing allows halftone and process art. The design is transferred to the object on a special machine that preheats the carrier and then presses it against the bottle with a heated roller. The bottle is oven- or flame-cured to improve adhesion and to add gloss. The system is especially effective on plastic bottles when decorating speeds are geared for high-volume production rates.

With quick-dry pigments, the need for 40-ft-long drying ovens previously required is eliminated. This had reduced the space necessary by over 80%.

Recent development include a special stretch-shrink carrier that can be used to decorate irregularly shaped bottles, such as tapered designs, without distorting the decoration. In addition, clear-release coatings have been developed that give the process widespread application with clear PVC and PS containers. Further developments have simplified changeover through the use of adjustable or universal-type tooling.

Di-Na-Cal, a development of Diamond International Corp., is a heat-transfer decoration system that allows simultaneous application of multicolor printing on two sides of a part. Up to six colors can be used on each side including process art and halftones in unlimited color ranges. Wraparound coverage is not possible; the maximum decorating area on cylindrical bottles is approximately 135°C per side, or about 75% of total coverage. Maximum advantage is obtained when two different designs are required (front and back) on a flat or oval part.

When printing on plastic bottles, the parts are internally pressurized by compressed air to provide the support necessary for transfer. This internal air pressure can also be varied to compensate for bottle deformities such as concavities, that may exist.

The process can readily handle odd-shaped parts with recessed areas; for areas not normally accessible by other methods, the internal air pressure can be increased during transfer to blow out these sections and create an area sufficiently flat for decorating. Another advantage of the system is its ability to handle extremely thin-walled containers. The continuous application of internal air pressure, plus controlled preheating, prevents damage to the container surface that would normally occur by other decorating processes.

Recent modifications have substantially increased production rates and allow simultaneous transfer to two or more containers in one operation. Multiple holding fixtures are used that stack the containers vertically, one above the other; the stroke of the transfer roller has been lengthened so that it can press the carrier web against several vertically stacked containers. This will have maximum effectiveness with smaller-sized bottles (under 8 ounce), where production volume can be substantially increased.

The Electrocal hot-stamping transfer is a system offered by Noble and Westbrook Div., The Bristol Brass Corp., and is a combination of hot-stamping, screening, and transfer-decorating. The design is screen-printed on a paper or plastic film carrier in roll form. Transfer of the design from the release-coated paper to the plastic part is accomplished with regular hot-stamping equipment modified to handle the special paper transfer roll. Transfer takes place in one standard hot-stamping cycle, using heat and pressure and silicone rubber-stamping dies that conform to the surface to be decorated. Abrasion-resistance and adhesion are good.

Electrocal has not been used effectively on blown plastics bottles, mainly because there is no effective method of providing internal support required to compensate for the pressure and heat involved. Also, the system is not so effective on PE as it is on more rigid materials, such as PS.

The Letraset transfer system, developed by Letraset, Ltd., London, England, is catching on as a means for decorating plastic bottles. In this process, transfers are silk-screened or gravure-printed on a clear carrier film and supplied in web form, ready to apply. The transfers are applied by conventional hot-stamping equipment modified by replacing metal dies with an aluminum-backed silicone pad.

This method allows multicolor decoration of either silk-screened or rotogravure designs in one pass and offers up to six-color printing. Silk-screened graphics can be used for short runs. While it is available for PVC containers, the process is not yet available for PE or PP blow-molded containers.

IMPORTANT CONSIDERATIONS

A decoration must remain attached to the container or product throughout its expected life. Various factors can affect adhesion; consequently, tests have been developed to simulate lifetime use of the product.

The pressure sensitive-tape is generally accepted in the industry as a preliminary test of adhesion. Crosshatches are cut on the decoration surface and transparent, pressure-sensitive tape is firmly placed over the hatching. The tape is then quickly peeled from the part. If large areas of decoration are removed with the tape, adhesion is poor.

The adhesion of paints or inks to plastic depends on intermolecular attraction. Each material presents a distinct decorating problem and pigments must be specially tailored for the specific type of plastic used.

Often a decoration will deteriorate or lose adhesion when it comes in contact with a product and pigments will react negatively with different products. Detergents, perfume oils, and alcohol will dissolve many types of pigments. In most cases the supplier has or can develop special pigments that are compatible with the product; in all cases, however, compatibility tests must be conducted before final approval.

A simple test includes immersion in the product over a predetermined period, for at least 24 hr, but it will depend on the amount of time the decoration might reasonably be in contact with the product.

Many plastics molders have been faced with a dilemma because they failed to consider decoration in advance or as part of the total package development program. Sometimes an entire design concept is destroyed and aesthetics sacrificed because essential decorating elements had to be omitted. Too often, decorating becomes a subject for consideration after the molds for the part are completed and production has started.

Many decoration problems can be avoided through slight modifications that do not greatly affect the total design concept. In other cases, special tooling can be developed to accommodate the object, if sufficient development time is allowed. There are also many established decorating suppliers available who can provide valuable services on a devlopment project. Such assistance can help eliminate unnecessary retooling at a later date.

Even modifications to a plastic formula will often require a completely different pigment formulation. Rigid, nonplasticized PVC, for example, requires different pigments than does highly plasticized PVC.

The rigidity, flexibility, and wall thickness of a container also have a direct influence on the decorating method used. As has been indicated, supporting devices and holding fixtures are often necessary to maintain the surfaces in proper position for accurate registration and to prevent damage to the container.

The wide range of designs, shapes, and styles allowed by plastics create other considerations. Whereas glass and metal containers are mostly cylindrical, plastics offer a variety of shapes and designs, including reverse-tapered, oval, and concave. Each presents distinct handling requirements that often dictate the type of decorating media to be used.

Polystyrene is a rigid but somewhat brittle material. It has a good polar structure and is solvent-sensitive to many pigments and inks, which allows it to be readily decorated by every known process with good adhesion. In the case of hot-stamping and Electrocal transfer decorating, sufficient wall thickness must be designed into the part and holding fixtures must be used so that the object can withstand the pressures involved.

Polyethylene and PP must be pretreated for all decorating methods except hot stamping. Surface pretreatment involves either flame or electronic heating to oxidize the surface and give it intermolecular attraction for pigments or inks. These materials can range from extremely soft and flexible to semirigid. Containers made from them are decorated by almost every process, with the possible exception of Electrocal.

The PVC used for blow-molded bottles is a semirigid material. It has relatively low heat resistance; consequently, hot stamping is extremely difficult because of the heat applied under pressure.

Containers made from this material can be effectively decorated by almost every other process with the possible exception of some transfer methods.

DYEING

Dyeing plastic parts after they have been molded or fabricated is a coloring method that is particularly valuable in family molding or where the molded part fails to meet color specifications. An interesting application of this technique is in the tinting of metallized and lacquered plastic parts.

The methods of dyeing plastics are essentially the same as those used for textiles. Two procedures of particular importance for plastics are bath dyeing and pad dyeing. The bath-dyeing method is most common for small molded items such as buttons, snap fasteners, decorative fringes, screws, telephone connectors, etc. The plastic parts are dipped into the dye bath for periods of time ranging from one-half to several hours. The dye bath is heated and constantly or intermittantly stirred. Some plastic materials require higher than 100°C temperature for satisfactory color buildup, in which case the dyeing is done under pressure. The padding/fixing method of dyeing is more suited to the surface coloration of foils and films. The item is dipped into a concentrated solution or dispersion of the dye. The excessive dye liquor is squeezed out and the substrate is heated in an oven to diffuse the dye into the plastic material. In both cases, the dyed goods are washed with a mild scouring agent and, if need be, the dyestuff fixed in the plastic material with the help of certain chemicals to enhance its washfastness properties.

There are basically three types of dyes which come under consideration for the dyeing of plastics: acid dyes, basic dyes, and disperse dyes.

The most commonly used acid dyes are those which are suitable for dyeing nylon fibers. The 1:2 metal complex dyes are of particular interest among this class of dyes: they have a high affinity for some plastics, especially nylons; their lightfastness is excellent; their solvent resistance is good; their washfastness is satisfactory.

Basic dyes are those which are used for dyeing acrylic fibers. These are known for their brilliance and high intensity of color, but their affinity for most of the common plastics is marginal. Their washfastness and lightfastness properties are not quite satisfactory. Their use is mostly in tinting of transparent plastics.

Disperse dyes are extensively used in synthetic fibers such as polyester, nylon, and cellulose acetates. They have a good affinity for most of the common plastics, the color build-up is generally satisfactory and the washfastness is excellent. Their solvent fastness properties are not as good as those of acid and basic dyes. The lightfastness ranges from excellent to poor, depending on the chemical structure of the individual dye. While the acid and the basic dyes are soluble in water, the disperse dyes are only sparingly soluble in hot water. They are therefore applied from nonaqueous solutions or from water dispersions.

All the dyeing processes, those used for textiles or those used for plastics, are dependent on the phenomenon of diffusion of dye molecules into the substrate for their dyeing action. The diffusion of dyes in the plastic material is influenced by the type of dye, the physical condition of the dyeing surface, the formulation of dye bath, the duration of dyeing time, and the temperature of the dye bath. Plastics additives, such as plasticizers, internal lubricants, antistats, etc., also affect the dyeing process and influence the fastness properties of the dye on the plastic part.

There is no standard procedure of dyeing plastics: each plastic composition has its own peculiarities. The formulation of dye bath and dyeing conditions must take into account these peculiarities as well as such criteria as the desired depth of shade, fastness properties and, of course, the cost of coloring. The dyestuffs suppliers can be of considerable help in the selection of dyes and the general guidelines for dyeing procedures. Test dyeing and assessment of fastness properties to light, water and solvents are the first necessary steps toward the development of any dyeing or tinting procedure.

Tinting or shading of metallized or lacquered parts is of considerable importance in plastics finishing. The molded plastic part is at first treated with a glossy base lacquer and dried. It is then vacuum-metallized. The metallized part is sprayed with a protective colorless lacquer, then the lacquered and dried metallized part is tinted in a dyebath, washed, dried and, if need be, protected with an application of glossy water-clear lacquer. Following this procedure, several different gold tones and other metal effects can be achieved. The technique of dyeing the metallized and lacquered part (instead of applying a colored lacquer coat) is necessitated by the fact that dyeing avoids the color-intensity variations caused by dripping or tear-dropping of lacquer. If the color is mixed with the lacquer itself, the color intensity will be proportional to the thickness of the coating. Then, if drips occur, they will appear as color marks. In the case of surface dyeing, on the other hand, the intensity of the color is uniform throughout the surface, as the dye has diffused through the lacquer irrespective of its thickness variation on the plastic surface.

1. Hall, Allen "Painting Plastics: the Alternatives to Solvents." *Plast. World* (Jan. 20, 1975).
2. Skeels, A. H. "Printing." *Modern Plastics Encyclopedia.* New York: McGraw Hill (1975).
3. Ayres, R. L. "Painting High Density Polyethylene." SPE Annual Technical Conference (1976).
4. Kut, S. "Painting." *Modern Plastics Encyclopedia.* New York: McGraw Hill (1975).
5. "Decorating Plastics 1980." *Plast. Eng.* p. 65 (Sept. 1977).
6. Conklin, H. J. "Pigments in Hot Stamping Foils." SPE Annual Technical Conference (1977).
7. Frazier, J. F. "Printing and Decorating Plastics." SPE Annual Technical Conference (1978).
8. Glatt, J. E. "The 'Sciem' Process." SPE Annual Technical Conference (1978).

10
Regulatory Requirements

INTRODUCTION

There are many federal, state, and local laws and regulations governing the use of colored plastics. The laws of interest to the plastics colorist are administered by the Food and Drug Administration (FDA) of the Department of Health, Education and Welfare, and the Animal and Plant Health Inspection Service of the Department of Agriculture (USDA). The three most relevant FDA statutes are the Federal Food, Drug, and Cosmetic Act, as amended, which relates to the use of colored plastics in contact with foods, drugs, and cosmetics, as well as the use of pigmented plastic devices; the Federal Hazardous Substances Act, which covers certain hazardous household substances and embodies the Child Protection and Toy Safety Act concerning toys and other articles used by children; and perhaps the Poison Prevention Packaging Act of 1970, which pertains to special packaging requirements of certain chemicals, drugs, and other items to prevent children from opening the containers or obtaining a toxic or harmful amount of the product. Two major laws under the jurisdiction of the USDA which relate to colored plastics are the Meat Inspection Act and the Poultry Inspection Act; they are concerned with plastics in contact with meat and poultry products in federally inspected establishments. In addition to the FDA and USDA regulations, there are potential regulations by OSHA covering occupational hygiene and EPA covering air-, water-, and soil-pollution by colorants and by solvent emissions in painting of plastics. No specific laws and regulations concerning colored plastics have as yet emerged from these agencies. Several studies are underway and the restriction that may emerge remain speculative. In 22 states, solvent-emission regulations are now effective. They are applicable to the plastics industry as well as to the metal and wood finishing industries.

COLORANTS AND THE FDA

Compliance with the FDA laws and regulations is a most important, complex, and also a most confusing area for the plastics industry. The industry, of course, is mainly interested in colored plastics intended for use in contact with food, not colors added directly to food (they are covered by the Color Additives Amendment to the Food, Drug, and Cosmetic Act).

With the signing of the Color Additives Amendment (not to be confused with the Food Additives Amendment) in July 1960, several changes have taken place in the law which are important to health and safety of the population. The ever increasing use of plastics for food wraps and containers and the coloring of them for aesthetic and sales appeal, have initiated many questions for the fabrication of these products. No absolute answers to these questions are available, but background information about these regulations should help the plastic colorist avoid any potential pitfalls and enable him to make intelligent decisions.

To place the Color Additives Amendment in its proper perspective, a review of the legislation on the subject is necessary. In 1906, the original U. S. Food and Drug Act was signed into law. This provided for a voluntary system of certification of the purity and safety of dyes used in or on food. This act was revised, to become the Food, Drug and Cosmetic Act of 1938. Under its provisions a list of colors was promulgated for use in three categories: Foods, Drugs, and Cosmetics; Drugs and Cosmetics; and the external application of Drugs and Cosmetics.

Included in this act were limits on the amounts of lead, arsenic, and heavy metals allowed in the colors, plus a provision setting down the specifications of each individual color. Manufacturers of the listed colors were required to submit representative samples of each color batch to the FDA for certification of the batch for the specified use as established in the regulations. The colors covered in this act were the so-called coal-tar colors: those derived from coal tar or a substance so related in chemical structure as to be capable of being derived from coal tar. This act did not include the nonaniline colors such as iron oxides, titanium dioxide, or the naturally occurring colors.

In 1958, Public Law 85–929 was passed by Congress. This is the Food Additive Act, which includes colors not listed in the 1938 act. The colors listed in the 1938 act were exempt from the Food Additive Amendment because of the prior sanction.

In July 1960, the Color Additive Amendment was enacted. This embraces all colors, both natural and synthetic, used in foods, drugs, and cosmetics. This new law, the Hill-Harris Color Additive Amendment (Public Law 86–618), was designed to replace inconsistencies and outmoded provisions which governed the use of colors in the 1938 act. An important additional purpose was to institute tolerance limitations when required and provide for other conditions of safe use. Official certification to assure safety to the consumer was continued. The new law changed the old in several respects. It did away with the difference in legal requirements and treatment between the coal-tar colors and other color additives, and it covered all color additives, natural, inorganic, and organic. It removed the unworkable harmless *per se* concept and replaced it with the requirements that the colors be safe under their intended conditions of use. It provided for separate listings of color additives for use in or on foods, drugs, and cosmetics. It provided for listing and certification of batches required under the old law but also permitted the Secretary of Health, Education, and Welfare to grant exemption from the requirements of certification when such certification is not necessary to the protection of public health. The Color Additives Amendment also contains an anticancer clause similar to the one in the Food Additives Amendment (the Delaney Clause) which states that any color which is capable of producing cancer in man or animal automatically is barred from use. However, legislative history will show that congress intended that a "rule of reason" be applied and the strictest interpretation not be made.

FOOD-CONTACT PACKAGING

The current status of the regulations and restrictions governing the use of colors for plastic food packaging are somewhat hazy. It would be reasonable for the industry to ask what colorants are permitted for coloring plastics to be used in contact with foods, and expect to receive a list of approved colors. At the present time, however, no clear rulings have been established concerning the use of colorants for plastics to be used in contact with food. No such listing of colorants is then possible. The regulations and restrictions concerning the packaging of food, drugs, and cosmetics in plastic containers and film emerge from other established regulations which permit the use of certain colors in other food, drug, and cosmetic packaging applications, and one or two regulations on specific colorants which include their use in certain plastics. There also are a few other colorants which have been exempted from food additive status because of their long history of safe use. Additionally, some food colors certified for use in food may be used in food-contact plastics.

The following four groups of colors are currently permitted in the coloring of plastics for food, drugs, and cosmetics.

1. Certified colors: a complete list of FD & C certified dyes and alumina lakes is available. The dyes are used for the coloration of such items as nylon. The alumina hydrate lakes are available in several forms; the so-called regular type, which has varying quantities of moisture and specially high-dried materials where the total moisture content has been reduced to a maximum of 5%. These colorants may be used in the various polymers and incorporated by the dry-blending process.

Additionally, they may be used in vinyl film and incorporated via dispersion in the plasticizer. If opacity is required, the use of purified titanium dioxide is recommended. In certain cases zinc stearate had been found helpful as an aid to dispersion, especially in polyethylene and polypropylene. However, the zinc stearate must be made from chickedema free-tested stearic acid.

2. Purified nonaniline colors, including titanium dioxide and carbon black, may also be used in and on plastics.

3. Petition may be made for the use of noncertified colorants for plastics. The burden of proof that the color does not bleed, crock, migrate or become part of the food, drug, or cosmetic, rests with the user.

4. As long as there is an impermeable barrier between the unapproved colorant and the food, drug, or cosmetic and there is no chance of the colorant becoming a part of the product either by design or accident, then the colorant is not subject to the color additive amendment. This means that any pigment or dyestuff may be used under these circumstances, the only restriction being the choice of color, depending on the particular usage and property of the colorant.

According to the food additive regulations, food additives include all substances not exempt by the Act, the intended use of which results or may reasonably be expected to result, directly or indirectly, either in their becoming a component of food or otherwise affecting the characteristics of food. A material used in the production of containers and packages is subject to the definition if it may reasonably be expected to become a component, directly or indirectly, of food packaged in the container. The regulations also state that if there is no migration of a packaging component from the package to the food, it does not become a component of the food and, thus, is not a food additive. Therefore, the packaging component is exempt from the law, and a food additive regulation is not required. This interpretation is the well-known "no migration" concept which has been applied frequently to the use of certain ingredients of food-contact articles, thus permitting their use without a specific regulation. This no-migration concept was reaffirmed by the FDA in a letter to the general counsel of the Society of the Plastics Industry (SPI), "where the review of the extraction studies confirms a lack of potential for migration, no regulation is necessary."

The majority of the colors used are quite inert chemically and extremely insoluble. Also, they are usually used in small concentrations, so the quantity which might contact food is low. These factors, plus the fact that the color may be protected from food contact by encapsulating or coating, almost assure little or no potential migration. In fact, this has been a major basis for the safe use of many colorants in packaging materials. There are two major difficulties to this approach, however. The list is strictly commercial in that it consists of convincing the user that there is no food additive problem or that your product complies with the law. Usually he will prefer to see a published regulation or, at least, a letter permitting such use. The second, more technical and legal, difficulty is the decision as to what comprises "no reasonable likelihood of migration to food." There is no definitive answer. Each situation may be different because of formulation, use conditions, solubility of colorant, etc. Many have used a value of less than 10 parts per billion and not diluted, but this is an arbitrary figure.

In addition to establishing a cutoff point, there is the obvious problem of analytical sensitivity. Inorganic pigments are metallic compounds or salts, and the extractives with food-simulating solvents may be analyzed for the metal involved by several different procedures. The heavy metal pigments, such as those based on cadmium, mercury, selenium, and lead, are of most concern because of the known toxicity of these compounds and their widespread use in plastics. Lead pigments do not find much use in food-contact materials, but several other metal compounds have been used, probably because of their lack of migration. Extraction tests with plastics containing these compounds must have adequate sensitivity to demonstrate a lack of migration. For organic pigments, analysts often rely on some type of color test to evaluate extraction. Most of these colorants possess such intense coloring ability at useful levels that extremely small quantities may be determined in food-simulating solvents.

CONTAINERS FOR DRUGS AND COSMETICS

Colored containers for drugs and cosmetics also come under the jurisdiction of the Food, Drug, and Cosmetic Act, but under different sections of the law, and they are regulated differently. There are no published lists of permitted container ingredients for food-contact materials. All drugs require FDA approval prior to marketing, and the container, which is considered part of the finished drug, must be approved along with the drug. Therefore, information on extraction and safety of the colorants used in the container is important. For cosmetics, preclearance prior to marketing is not required for food additives and new drugs, but the finished cosmetic must be safe for its intended use. The container must not be composed of any substance which renders the content harmful. Extraction and safety data would be helpful to the cosmetic manufacturer in determining the safety of his product.

USDA REGULATIONS FOR MEAT AND POULTRY CONTACT

The USDA has issued the bulletin on "Resinous Plastics as Indirect Additives." This publication has a section on colorants which reads:

> Colorants considered acceptable include: titanium dioxide, carbon black, iron oxides, phthalocyanine blue and green, Prussian blue, lakes of FD and C dyes, and certain organic coloring materials. The colors must not migrate to the product nor cause a deceptive appearance. Salts of metals known to form toxic compounds such as arsenic cadmium, chromium, lead and mercury are not acceptable.

This USDA bulletin concerns containers, films, bags, trays, formed resins, and other plastics which accidentally or otherwise come into contact with meat and poultry food products produced under Federal inspection.

OCCUPATIONAL AND ECOLOGICAL CONSIDERATION

Beyond FDA and USDA jurisdictions, regulations for colorants become speculative. Some further restrictions are possible, even probable. These are the results of concern by OSHA about occupational hygiene and EPA about possible air-, water-, and soil-pollution by colorants, especially those based on heavy metals.

At the moment, widespread responses to OSHA's activities have included extensive and detailed inventories of information about the safety of colorants and associated raw materials in printing ink, coatings, and plastics. These efforts continue even in the absence of any specific concern by OSHA for occupational hygiene related to colorants in plastics. Regardless, it would be well for all concerned to review and improve internal procedure for protecting workers exposed to dusts of colorants. It is true that colorants in plastics are extremely low in OSHA's priority, but this does not justify complacence about the development of applicable standards and inspections to determine compliance. OSHA has appointed several standards advisory committees to develop job safety and health regulations for a variety of materials, some of which are related to colorants. These include silica, lead, arsenic, cadmium, and mercury.

Probably of more immediate interest is the concern of OSHA, EPA, and also NSF for air and water pollution by trace amounts of heavy metals from the manufacturing and disposal practices, explicitly including colored plastics. EPA directs its attention to the welfare of the general public. The NSF serves as a source of new information, on these and other subjects, by providing funds to various university and private research organizations proposing relevant projects under the Research Applied to National Needs program.

Cleaning coloring equipment, worker's hands, and protective gears, in the interest of good occupational hygiene, could lead to water-pollution problems of concern to EPA. The level of pollution caused by incineration of municipal waste, especially plastics, is also of real current concern to EPA and to NSF's Division of Environmental Services. The impact of dissipated heavy elements, especially cadmium and mercury, on the environment, are of particular interest to both the NSF and EPA studies. It is estimated that over 200,000 lb of cadmium (as metal) is generated by incineration of pigments and stabilizers in plastics. A recent report indicates that weathering releases 230 metric tons of mercury per year – a considerable amount for an ecologically minded population.

The above comments by no means imply that colored plastics are a major source of water- and air-pollution, or that the plastic industry is among the target industries identified by the Labor Department and EPA. While the danger of pollution is real, the current level of plastics-produced pollution permit restrictions to remain speculative.

In addition to the areas of concern mentioned above, EPA touches the plastics industry in another indirect way. To the plastics processor who paints plastic parts and has plants located in any of the 22 states that now have solvent-emission regulations, as of May 31, 1975, he had to comply with those regulations or face sanctions by the EPA under the Clean Air Act. All but two of the states with solvent-emission regulations have based their requirements on Los Angeles' landmark Rule 66. What that rule did was to arrange commonly used solvents in order of photoreactivity – hence, severity of their effects in the atmosphere – and establish limits on the amounts that could be released into the air on a daily and hourly basis.

Solvents that have low reactivity have been exempted from the ruling and can still be used in paint formulations under regulations based on Rule 66. However, use on exempt solvents (or Rule 66 solvents, as they have come to be called), is permitted only with air dry-type coatings. If bake-dry coatings are used, the amount of any solvent released into the air must be limited to 15 lb per day or 3 lb per hr.

In the case of the reactive or restricted solvents, Rule 66 allows for the release of 40 lb per day or 8 lb per hr, an amount that is equivalent to only about 15 to 18 gal. of paint per day.

BIBLIOGRAPHY

1. Dunn, M. J. "Colors for Food Packaging and FDA Regulations." SPE Annual Technical Conference (1965).
2. Dunn, M. J. "Colors for Food Packaging." SPE Annual Technical Conference (1968).
3. "The Law and You." *Plast. Eng.* p. 59 (Aug. 1977).
4. Ingle, G. W. "Non-FDA Restrictions on Colorants for Plastics." SPE Regional Technical Conference, Philadelphia (Oct. 1972).
5. Miller, R. M. "Regulatory Compliance of Colored Plastics." SPE Regional Technical Conference, Philadelphia (Oct. 1972).

11
Mass Coloration of Man-Made Fibers

INTRODUCTION

Sandwiched between the two large sectors of textile dyeing and plastics coloration is an interesting field covering the mass coloration or spin-coloration (sometimes also called dope-dyeing) of man-made fibers. Although the man-made fibers are actually the starting products of the textile industry, for all practical aspects the problems of mass coloration are related to those of coloration of plastics.

The early work in the field of mass coloration of man-made fibers dates back to the time when the fibers themselves were being developed. Although at the time the reason for conducting investigations was to be found in the lack of suitable colorants and methods of dyeing, the factors now giving mass coloration a fresh importance are economic considerations, problems of staffing, and problems related to the environment.

The world output of man-made fibers in 1977 was approximately 10 million tons, comprised of cellulosics (rayons and acetates), acrylics, nylons, polyester, and polypropylene. It is estimated that about 5% cellulosics, 3% acrylics, 2% nylons, 1% polyester, and in excess of 50% polypropylene fibers are colored through mass coloration. Assuming a mean concentration of 1% colorant, calculated as dry colorant on weight of fiber, this market represents a substantial consumer of colorants, comprising pigments, pigment dispersions,

Table 13. Estimated world-wide consumption of colorants for mass-coloration of synthetic fibers (1977).

Fiber	Total Production (tons)	Colored Fibers (tons)	Colorants Consumed (tons)
Acrylics	1,732,000	52,000	520
Nylon	2,830,000	57,000	570
PET	3,907,000	39,000	390
Polypropylene	900,000	45,000	520
Total	9,369,000	193,000	2,000

and "oil soluble" dyes. The various estimates are summarized in Table 13. The probability is that the figures are conservative. In any case they justify the premise that mass coloration of synthetic fibers is a business of major proportions. It is not surprising that all the major colorant manufacturers offer products specifically designed for mass coloration of synthetic fibers, or that the patent literature contains numerous references to their endeavors in this field.

Somewhat singular is the use of a plethora of names for this method of coloring which has persisted, often erroneously, in all forms of writing on the subject, e.g., "spun-dyed," "mass-dyed," "spin-colored," "solution-dyed," "dope-dyed," "pigmented," etc. The emergence of "mass coloration" as the recommended generic term is welcomed. It should be universally accepted together with the definition: A method of coloring a man-made fiber by incorporation of the colorant before the filament is formed.

The origins of mass coloration have been recounted by several authors, including Ackroyd (see Bibliography) who gives valuable references to earlier works on the subject. The mass-colored viscose was well established by the early 1950s. The earliest published reference to the availability of mass-colored viscose continuous filament yarn and staple fiber appears in a lecture by Hegan given in October, 1943 to the Rayon Converters section of the Cotton and Rayon Merchants Association (England) as one of a series subsequently printed in book form for private circulation under the title "Talks on Rayon and Synthetic Fibers." It was not until 1953 that production of man-colored viscose rayon started in the United States.

Since viscose was the first man-made fiber to be exploited commercially on a large scale, clearly the early history of mass coloration is the story of the mass coloration of viscose. All the historians commend it as a bright idea, but, curiously enough, none spells out what must have been a constant spur to innovation, namely, the need to overcome the effects of acutely variable dye substantivity which afflicted viscose rayon in its early years.

The question is often asked, What are the advantages of mass coloration? Or, conversely, What are the disadvantages of mass coloration? The advantage of uniform coloration and excellent all-round fastness, together with fiber quality equal to that of the uncolored fiber, since no additional processing is entailed, are the points on the plus side. The lack of flexibility in production and the necessity of maintaining large inventories, are the points on the minus side.

A cost comparison between spin coloration and conventional coloration (dyeing) in absolute figures is virtually impossible owing to the large number of different fibers, colorant ranges, and coloration processes. A few points can, however, be mentioned: the colorant costs in conventional dyeing are about 100% higher than in mass coloration; the energy costs, which are an important cost factor in conventional dyeing, can be ignored in mass coloration; the cost of spin finishes and other textile auxiliaries are nil in mass coloration; wage costs in conventional dyeing are about ten times as great as in mass coloration; contamination of waste water by alkalies, acids, bleaching agents, surfactants and carriers depends on the particular process employed, but is always a very substantial cost factor in textile dyeing, whereas in mass coloration it entails no cost at all. On the negative side, material consumed in the matching and changing of shades and that wasted in possible machine down times are cost items unknown in conventional dyeing.

CELLULOSIC FIBERS

Mass coloration of viscose rayon was found, in real terms, on the successful bulk production of aqueous dispersions of carbon black. Other pigment dispersions which followed in rapid succession were based on C.I. Pigment Red 7, C.I. Pigment Yellow 1, and C.I. Pigment Red 2. These are all azo pigments found to withstand the conditions of viscose rayon production, and C.I. Pigment Blue 15 (copper phthalocyanine) was an obvious candidate used in aqueous dispersion for viscose at about the same time.

Initially, viscose rayon was mass-colored using a *single-batch discontinuous system.* Continuous production was made possible by the development of the *masterbatch system,* in which a quantity of viscose is mixed with pigment paste to give a concentration several times that required in the viscose for spinning and this mixture, i.e., the masterbatch, is injected, in metered amounts using a standard viscose two-year pump, into a blender fed with metered uncolored viscose from the main supply. Refinements of the masterbatch system followed with the introduction of a highly efficient homogeniser and devices to reduce the human element in the laborious repeated preparation of masterbatches. Slack and Parr Ltd. developed their Fluid Flow mixer with high-shearing and multiple-transfer actions, and also the three-gear pump. The mixer provides very efficient blending with minimal heating of the dope. The three-gear pump, with two inlets and a common outlet, meters directly to the homogeniser a 1:1 mixture of diluted aqueous pigment dispersion and main-supply uncolored viscose to produce, in effect, a continuous masterbatch. The invention of AKU, frequently referred to as injection by differential pumping, utilizes two two-gear pumps to perform the same function. Elegant automatic control systems were the next development, giving constant pigment/viscose mixtures irrespective of fluctuating demands at the spinning machine. Introduction of an improved version of the high-efficiency mixer by Slack and Parr, together with precision piston and flexible diaphragm pumps capable of metering dilute pigment dispersions of water-like viscosity accurately against line-pressure viscose at 10 psi, enabled masterbatches to be dispensed with altogether and the combination became known as the *direct injection system.* It has now been widely adopted for all scales of production.

Once correctly set-up, the system ensures maximum efficiency and colorant utilization, minimal labor costs and colorant wastage, and excellent flexibility for color changes and adjustment. Long runs of uniform material in all intensities from the palest pastels to maximum depth can be accommodated by appropriate variation of the concentration of the diluted pigment dispersion, maintaining, meanwhile, a constant viscose: pigment slurry ratio at the metering pumps. Another benefit is the minimal adverse effect on the textile characteristics, viz. decitex, tenacity, extensibility, etc., of the colored fiber.

A most important factor in the achievement of the direct injection system was steady improvement in the physical form of pigment dispersions by the colorant manufacturers. In particular, mean particle size was progressively reduced by a variety of comminution techniques often specific to the different pigment types. A disc centrifuge developed commercially by Joyce, Loeble & Co. Ltd., Gateshead, England, under license from Imperial Chemical Industries Ltd., has proved a valuable tool for process development and quality control. The *particle-size disk centrifuge* does not require skilled operators and is suitable for routine use, but is, nevertheless, an instrument capable of giving precise analyses of particles in the range 0.01–30 μm equivalent Stokes diameter.

Developments so far considered have been in techniques and equipment, but the relevant field of color chemistry has also been well tilled. The seven pigments adopted in the earliest stages remain, nevertheless, among the "bread and butter" products of the industry. Certain of these pigments result in marked opacity (sometimes referred to as "blinding") of the filaments, which is often an advantage for staple fiber but is generally undesirable for continuous-filament yarn. Later work pointed to the reasons for opacity and dichroic effects. These factors were taken into account when the original seven were increased, primarily to widen the color gamut and to overcome fastness deficiencies in some cases.

In due course a range of second-generation dispersions became widely available based on, in the main, C.I. Pigment Yellow 49, C.I. Vat Orange 9, C.I. Pigment Brown 1, C.I. Pigment Red, 5, C.I. Pigment Red 11, C.I. Vat Violet 1, C.I. Vat Blue 1, C.I. Vat Blue 9, and C.I. Pigment Green 7. Specialities followed; for example, a vogue for bright bluish maroons and bordeaux colors requiring excellent light and dry-cleaning fastness was met; a phthalocyanine green was produced having a much reduced dichroic tendency, as were a particularly intense homogeneous brown, C.I. Pigment Brown 22, a violet with outstanding fastness properties, C.I. Pigment Violet 23, reddish transparent oranges based on C.I. Vat Orange 7 and C.I. Pigment Red 168, and many others.

A certain penalty accompanied these advances. Spinneret incrustation, the gradual blocking of spinneret holes, occurred especially with vat-type pigments as a result of partial solubilisation of superfine particles by the reducing action of the viscose solution and subsequent precipitation at the spinneret spin-bath interface. Sodium m-nitrobenzene supphonate suggested as a reduction retarder and non-ionic, isotonic pigment dispersions nonflocculatable by viscose were developed. In the normal course of development, these, and other spinning difficulties associated with mass coloration, resulted in numerous patents claiming the use of additives, but the tendency is for these to have been overtaken by better understanding of mass-coloration requirements by pigment manufacturers and fiber producers. Means of overcoming the fault of "inflation" (abnormal hollow filaments) were suggested and of incrustation with copper phthalocyanine pigments, and dispersion improvers are typical.

Reference has already been made to the search for lustrous transparent mass-colored fibers, and this was given new impetus following the advent of reactive dyes for cellulose in early 1956. Very attractive bright lustrous colors can be obtained by incorporating reactive dyes into viscose, and several processes have been claimed, but no significant commercial production has emerged because only a very few dyes given adequate fixation and the fastness properties of the colorings are very much inferior to those of pigmented yarns.

It has long been popularly supposed that viscose is moribund, but, if it has not shown the phenomenal expansion of the synthetic-polymer fibers in the last few years, its producers have displayed remarkable resilience and ingenuity in creating new and desirable forms of regenerated cellulose. In the light of recent events the unique advantage of all the cellulosic man-made fibers is worth repeating, nemely, this is the only species for which the basic raw material is naturally reproducing. It is reasonable to assume, therefore, that viscose will retain its important position and may even increase in volume, if unspectacularly.

Mass coloration will continue likewise because the same resilience and inventiveness are still bearing fruit. Although it is understood that they are not at present commercially available, novel pigment preparations have been developed in recent years which give colorings that approach the trasparency of those obtainable with soluble dyes while still retaining the fastness properties of those given by pigments. Equipment is now on the market which overcomes to a large extent the lack of flexibility, especially in manufacturing continuous-filament yarn, by permitting simultaneous production of several different colors on one machine. Adaptability to changing fastness requirements is illustrated by the current availability of mass-colored staple fiber for blends with nylon and polyester without staining off on to adjacent synthetic polymer fiber during high-temperature processing. Freedom from staining of poly(vinyl chloride) coatings on mass-colored viscose fabrics has also been found to be possible by careful selection of pigments. Possibly of greatest importance are the signs of a growing, if belated, awareness by fiber producers of the considerable potential of mass coloration for the promotion of purpose-designed colorings for specific large-volume outlets with the attendant benefits of excellent all-round fastness and fiber quality at modest cost.

Systems corresponding to the single-batch discontinuous and the masterbatch continuous methods for viscose are used for cellulose acetate fibers. The direct injection system is technically feasible using solvent-soluble dyes with precision metering pumps, but it has not been adopted in commercial practice because of the limited solubility of the dyes and control problems arising from the use of volatile solvents. For the same reason there is a greater preference for the single-bath system for these dry-spun fibers than applies for the wet-spun viscose. Mass coloration of secondary cellulose acetate is virtually confined to continuous-filament yarn, but triacetate is mass colored in both staple and continuous-filament forms.

Injection pumps and homogenizers are similar to those used for viscose except for the construction materials. Whereas cast-iron and mild steel are suitable for handling viscose solution, stainless steel and bronze are used to counteract corrosion, more particularly when the cellulose ester solutions (dopes) are mass colored with solvent-soluble dyes.

Two broad classes of colorant have been established: pigments and solvent-soluble dyes.

A variety of methods have been used in the past to prepare dispersions of pigments in secondary cellulose acetate before their incorporation into the bulk of the dope, including ball-milling in dope diluted with acetone to reduce its viscosity for efficient milling, and plastic milling using a high-boiling point solvent. One of the most satisfactory methods to emerge is intensive plastic milling of commercially pure pigment and acetate flake with a plasticizer such as dimethyl phthalate on heated two-roll mills. Although rather time-consuming, excellent dispersions are obtained and the thin sheets of highly pigmented acetate produced, containing 30–50% colorant, are broken down into powder or fine granules to facilitate mixing into dope for single-batch mass coloration or into acetone (with dope if necessary) for the masterbatch system. Pigment preparations in this form are available from colorant manufacturers and are extensively used by the fiber producers for convenience and reliable color standardizing.

The chemical types of pigments range more widely than those for viscose, since they do not have to withstand the same chemical agents, but all must be insoluble in acetone. Pigments used in both fibers will be noted among the following which are representative of the pigment preparations available: C.I. Pigment Yellow 17, C.I. Vat Yellow 4, C.I. Pigment Yellow 3, C.I. Vat Orange 3, C.I. Vat Orange 7, C.I. Pigment Brown 22, C.I. Pigment Red 7, C.I. Pigment Red 2, C.I. Pigment Blue 15, C.I. Vat Blue 1, C.I. Pigment Green 7 and C.I. Pigment Black 7 (Carbon Black). Among the later developments from colorant manufacturers are preparations based on C.I. Pigment Red 144 and C.I. Pigment Blue 60.

Solvent-soluble dyes have been developed simultaneously with pigments, and their use has extended the color gamut, particularly in the yellow to red region. All give lustrous transparent colorings, but the fastness properties of the colorings do not in all cases match those obtained with pigments. Too numerous for individual mention, examples may be found throughout the Solvent Dye section of the *Color Index* from which information on more recent introductions and specialities can be gained. Chemical types include monoazo and metal-complex azo, anthraquinone, and phthalocyanine compounds, anionic, cationic, or nonionic. They are commercially pure dyes free from standardizing and dispersing agents. This partly accounts for their apparent limited solubility in acetone, which is generally of the order of 5% wt/wt, but complete solution is not essential provided the mixture is homogeneous and stable throughout the period of use. The critical factor is solubility of the dye in the dope at spinning concentration. Depending on the state of the dope at the coloration stage, whether it be secondary cellulose acetate in acetone with a little water, or tri-acetate in methylene chloride with a small proportion of methanol, further filtration is inadvisable, especially when the single-batch method is used, in order to maintain optimum uniformity of color throughout the batch of fiber.

Some quantities of mass-colored tri-acetate staple fibers are being produced by Celanese Canada through the melt-spinning process. The pigments are blended in the right proportions with the dope before flaking the acetate out. This is purely a batch-type operation.

Production of mass-colored continuous-filament yarns is already running at a much higher level, probably two to three times, that suggested for the cellulosic fibers as a whole. Such yarns have a secure niche in fashion fabrics for which the industry is well organized, and have proved ideal to withstand the partial saponification process, which has become widely adopted to enhance the handle of triacetate continuous-filament fabrics and confer antistatic properties. There appears to be no reason why production of the attractive and durable effect obtainable should not follow the lines suggested for viscose.

ACRYLICS

Whether or not variable dye substantivity was a spur to mass coloration of viscose, the reason for development of polyacrylonitrile fibers was, as for acetate, the lack of suitable dyes and dyeing processes. For acetate the problem was solved by the SRA colors, the forerunners of modern disperse dyes. For polyacrylonitrile the fiber was changed instead of the dyes and, by producing appropriate copolymers dyeable with acid or basic dyes, the fiber makers again provided the solution. They also established four different spinning systems of commercial significance. Three entailed spinning solutions of polymer in organic solvents with both wet and dry processes when using dimethylformamide (DMF) as solvent and wet spinning only for dimethylacetamide (DMA); in the fourth the fiber is wet spun from sodium thiocyanate solution.

Pigment preparations similar to those for acetate are used for the processes dependent on organic solvents, but these are frequently compounded by the fiber producer using the local polymer, as the patent literature shows. It is noteworthy that the preparations are not restricted to polyacrylonitrile solvents or polymer but in some cases use other compatible materials for greater flexibility of application. Procedure and equipment are generally similar to those for acetate, but care has to be exercised in the selection of pigments because DMF and DMA are particularly powerful solvents; modern high-grade pigments like the quinacridone and perylene types are particularly valuable because of thier high-solvent resistance.

Clearly, solvent-soluble dyes cannot be used for the two processes in which polymer in organic-solvent solution is wet spun since the polymer-dye solvent is present in the spinning bath and heavy color bleeding would occur. Neither can solvent-soluble dyes be used in dry spinning because the consequences of using hot spinning solutions (ca 150°C) and high ambient extrusion temperatures (ca 400°C) are rapid and severe spinneret incrustation.

The attractively transparent effects from soluble dyes can, however, be obtained in conjunction with the wet spinning process using solutions of inorganic salt. The method does not, strictly, come within the terms of mass coloration, as defined before. It should nevertheless, be mentioned as it is also exclusively, a producer-coloration technique. For many years regenerated-cellulose films have been colored by virtually instantaneously dyeing, by running the cast film through a solution of a direct dye and washing before drying (the cellulose is extremely receptive to direct dyes in its swollen state). The same principle has been adapted for acrylic filaments destined for staple fiber, wet-spun from sodium thiocyanate solution. The filaments after extrusion are washed in tow form and then dyed by a continuous process with modified basic dyes having the required solubility, and whose stability at low pH remains unaffected by low concentrations of sodium thiocyanate. After dyeing the tow is rinsed, stretched, and finally dried. Neochrome (Courtaulds) and Zefkrome (Dow-Badische Co.) are now well-known brands of acrylic fibers producer-colored by this means. The process has also been adopted for coloring fibers wet spun from an organic-solvent solution. Coloring of excellent fastness are obtainable, and the process approaches ideal in flexibility; rapid changes of color can be made, and relatively small lots can be dyed economically without the necessity of interrupting fiber extrusion or incurring undue wastage of colorant.

The fiber producer is offered the following products by the relevant pigment manufacturers:

Pure pigments, which have to be selected and dispersed by the processor himself to the generally applicable fineness for spin coloration of under 2 μm in a suitable medium such as the solvent and some polymer, the choice of pigment and dispersion conditions depend on the spinning process, solvent, price, and fastness properties. According to requirements, diaryls, flavanthrones, monoazos, perylenes, phthalocyanines, quinacridones and other pigments may be used.

Pigment concentrates, namely, 40–60% pigment dispersions in acrylic polymer, the pigments being selected for universal applicability in all the spinning processes ordinarily used in works practice. For application by the injection method, products of this type merely need to be stirred into the appropriate solvent until the carrier material is completely dissolved.

The latest development in this field is that of the diluent-free *cationic colorants,* which dissolve readily in dimethyl formamide and dimethyl acetamide, by far the most important solvents for polyacrylonitrile. When injected into the spinning solution they fix so rapidly and completely, forming salt with the acid groups of the polymer, that they are suitable for use not only in the dry spinning process but also in the wet spinning process.

The main criterion in the develpment of these colorants is that the entire quantity should be rapidly fixed, thus preventing coloration of the coagulation and washing baths used for the wet spinning plants. Other important criteria are high solubility in DMF and the lowest possible solubility in water. The advantages of using this process compared with the use of pigments are

- high tinting strength
- greater brilliance
- reasonable price
- no filter blockages
- no spinneret blockages
- no impairment of fiber properties

NYLON

The mass coloration of nylon is essentially restricted to continuous-filament yarns. Mass-colored nylon fibers are increasing in importance. On a world-wide basis, about 1% of nylon fibers are colored *en mass,* mostly in Europe. In the United States only two companies are engaged in mass coloration of nylon 6 fiber but in the USSR almost 25% of the production is colored through mass coloration. Unlike acetate or acrylic fibers there was no problem during initial development in finding dyes which would color nylon 6 or nylon 6.6 satisfactorily, but – perhaps with the difficulty of achieving high fastness and uniformity of color in mind – DuPont examined the possibilities of mass coloration almost from the outset of commerical production and had published patents covering the incorporation of pigments or delustrants, first in powder and then in aqueous slurry forms, into the nylon before polymerization as far back as 1940–1943.

This apparently promising process, however, had a number of shortcomings, and, although considerable effort has since been expended, worldwide, on research into colorants and processes for mass coloring nylon, no entirely sufficient single system or, indeed, combination of systems has yet been devised because of the few colorants that survive the combined effects of the strong reducing character of the nylon melt and the high temperatures of the molten polymers (nylon 6, ca 215°C; nylon 6.6, ca 250°C).

In addition to these adverse factors, hot monomer in the autoclave was found to promote flocculation of pigment slurries, and so incorporation techniques were modified, and protective colloids and additives other than dispersing agents suggested a reduction in the flocculation. Mass coloration using the autoclave was extended by melt-spinning mixtures of differently colored nylon chips to produce addition colors. Highly pigmented "master-batch" chips were also produced for mixing with uncolored chips and subsequent spinning, and both Du Pont and British Nylon Spinners claimed processes for simultaneous continuous-mass coloration and polymerization. Colorants first used in the autoclave were mostly inorganic and included cadmium pigments; oxides of iron and chromium; ultramarine, with copper phthalocyanine; and, of course, carbon black. Since the early days numerous heat-stable types of pigment have been suggested, notably phthalocyanine derivatives, anthraquinone derivatives, quinacridones, perinones, and many others. Mass coloration of high-melting point polymers is difficult, and the incorporation of colorants in the autoclave has declined considerably, not only because of the inconvenience and high cost of cleaning before changing the color but also on account of the degradation of the polymer and subsequent undesirable effects on yarn quality.

Polymer degradation was alleviated by the introduction of chip coating which reduced the time of contact of the colorant with molten polymer to that during the spinning process only and also eliminated autoclave contamination. In chip coating, colorants in one form or another are applied to the surfaces of the chips and, provided the coating is consistent and stable, the technique readily leads to a uniform distribution of colorant in the polymer without the necessity for special mixing procedures or carefully metered simultaneous streams of polymer and pigment to the melt pool. Production of a stable coating of colorants on the chips was assisted by mixing and controlled heating and by incorporating additives as colorant vehicles – extenders with adhesive effects such as paraffin wax, low-molecular weight polymers, softening agents and silicones.

Following experience with these types of chip-coating compositions it seemed desirable to have the colorant dispersion as an integral constituent of the extender particles. A convenient masterbatch form became a possibility when processes were invented in which the extender was chemically identical with the polymer to be pigmented. It was, therefore, necessary to dissolve the polymer before mixing it with the dispersed colorant, and then precipitating and coating the chips before spinning. Different methods suggested included the use of water-miscible liquid aliphatic acids, hot water-soluble alcohols, high-temperature grinding, and plastic milling with formic acid. An ingenious variation on the two themes for coating compositions was the process of ICI based on a stable aqueous dispersion composed of pigment particles and similarly fine particles of the fiber-forming polymer.

Although the benefits of flexibility and generally improved fiber quality resulting from chip coating as compared with autoclave coloration were recognized, so also were new disadvantages such as color variation arising from dusting off and the relatively large amounts of colorant vehicles required and their high cost. An approach to these problems was, therefore, made by using as colorants nylon dyes with the necessary heat stability, and attaching them to the chips simply by dyeing in aqueous dyebaths using processes and equipment somewhat similar to those used for dyeing nylon textiles. The idea of chip dyeing seems to have originated with Hoechst, for polyester chips, but it is now frequently used for nylon 6 and to a less extent for nylon 6.6. The rather more severe conditions in the melt tend to restrict the choice of dyes, as of pigments, for nylon 6.6, but, for nylon 6, it is to some extent a case of necessity becoming a virtue. Nylon 6 production requires aqueous monomer extraction as an essential step and it is convenient to combine dyeing of the chips with monomer extraction and the subsequent drying operations. Nylon 6.6 does not require monomer extraction.

Numerous metal-complex azo dyes have been claimed for chip dyeing as have anthraquinone dyes.

Melt spinning of chips mass colored in the autoclave at spinning strength, e.g., for black especially, requires no special equipment other than small mixers or mills to prepare the colorant dispersions for addition to the pressure vessel. The homogeneously colored chips are charged to a feed hopper, protected at all times from decomposition by a blanket of nitrogen. Formerly these then fell by gravity to a heating coil and were melted. The molten polymer dropped down to a melt pool and through a metering pump (a development of the viscose two-gear pump suitable for hot viscous liquids), to the spinneret (via a sand-pack) or a similarly heat-resistant filter. A certain amount of mixing took place at each stage, but the heating coil or grid has now been largely superseded by the screw-extruder, which among other advantages, is much more efficient as a blender and greatly assist production of uniformly colored yarn from all forms of mixed chips, e.g., uncolored chip with master-batch highly pigmented chip, or blends of differently colored chips. More recently, in-line static blenders (Kenics mixers) have been fitted between the screw-extruder and the metering pump to increase mixing efficiency still further. Careful chip preparation before loading to the feed hopper is, clearly, of primary importance, not least in relation to the design and operation of tumble mixers for batch blending, color mixing, and chip coating.

Table 14. Pigment Usefulness.

Good	Fair	Poor
C.I. Pigment Yellow 37	C.I. Pigment Red 122	C.I. Pigment Yellow 93
C.I. Pigment Yellow 110	C.I. Pigment Red 149	
C.I. Ciba-Geigy Yellow 4610	C.I. Pigment Red 177	
C.I. Pigment Orange 20	C.I. Pigment Violet 19	
C.I. Pigment Red 108	C.I. Pigment Violet 23	
C.I. Pigment Blue 15		
C.I. Pigment Blue 22		
C.I. Pigment Green 7		
C.I. Pigment Green 36		
C.I. Pigment Black 7		
C.I. Pigment White 6		

About 25% of the total nylon 6 production is presently being colored through a mass-coloration process in the USSR. In western Europe Hoechst and several other companies offer mass-colored nylon filament yarns. In the United States, DuPont offers a few basic colors. One company (Camac, Bristol, Va.) offers a full range of fashion colors in nylon 6 filament yarns for upholstery and carpet markets. It appears that most of the producers use predispersed pigments and utilize chip blending technique, common to plastics industry.

Table 14 lists many of those pigments found to have adequate thermal and chemical stability to render them useful in Nylon 6 melt spinning. The listing is by Color Index designation. Those not yet so designated are listed by manufacturer.

The ratings for usefulness, relating to color stability in molten Nylon 6 are based upon normal melt spinning temperatures of 260° to 285°C for maximum dwell periods of 3½ min.

Available also for polyamides are complete lines of totally and partially soluble "pigment" systems, designed for inclusion in the hot melt. These products vary in their degree of solubility and some predispersion work may be necessary. They have a distinct advantage in color stability in the melt spinning process.

Those pigments which survive the exposure to molten Nylon 6 do not appear to have significantly altered the physical properties of the fiber. However, in long-term exposure to light under standard Xenon arc Fade-O-meter conditions the pigments affect widely diverse

Table 15.

Pigment in Stabilized Nylon 6	Fiber Strength Retention 1000 Hr. Xenon Arc SFH
Control – unpigmented	53%
0.75% Pigment Blue 15	70%
0.35% Pigment Black 7	60%
0.75% Pigment Green 36	51%
0.75% Pigment Green 7	48%
0.20% Pigment Violet 23	38%
0.75% Ciba Geigy Yellow 4610	20%
0.75% Pigment Yellow 93	18%
0.75% Pigment Violet 19	17%
0.75% Pigment Yellow 110	14%

results in tensile strength and elongation retention. Some pigments, under these conditions, enhance stability while others prove to be detrimental. Table 15 classifies some of the pigment types according to their effect on a commercially available stabilized Nylon 6 after 1000 hr Fade-O-meter exposure. The fiber, intended for carpet yarns, has a filament size of 25 denier (denier = wt in grams/9000 meters) and a diameter of 56 μ.

Further experimentation with filaments one half the above diameters, 28 μ (6.25 denier), display up to 50% greater strength loss for the same exposure period; thus, one can begin to evaluate color fade characteristics. Color retention of 0.75% pigment for the 6.25-denier filaments is listed in Table 16, with pigment types listed in groups according to the length of exposure (Xenon Arc SFH) required to achieve a fade of significant value comparable to a "3" on the AATCC Grey Scale.

Pigment dispersions in a carrier other than Nylon 6 are common. Polypropylene, polyethylene, polyethylene-vinylacetate copolymers, ionomers, and nylon 11 are some of the carriers used for the preparation of color dispersions which are then blended with virgin nylon chips. All these systems have their advantages and disadvantages and one should weigh such questions as polymeric compatibility: Is it important? Is it necessary? Which polymer will prove to be the best pigment wetter? Which will develop the greatest shear in compounding the dispersion? Are there handling problems? Particularly in the case of a noncompatible system, the rheology of mixing in

Table 16.

Group I – 500 SFH Xenon-Arc
Pigment Yellow 110
Ciba Geigy Yellow 4610
Group II – 750 SFH Xenon-Arc
Pigment Red 149 (1% pigment)
Pigment Red 177
Pigment Violet 19
Pigment Yellow 93
Group III – Over 1000 SFH Xenon-Arc
Pigment Violent 23
Pigment Blue 15
Pigment Blue 22
Pigment Green 7
Pigment Green 36
Pigment Black 7
Pigment Yellow 37
Pigment Orange 20

the melt spinning process becomes of more dramatic importance in producing a uniform discontinuous polymer system in nylon 6.

POLYESTER

Mass-colored polyester fibers are available in both staple and continuous-filament form; staple fiber being the more important. Poly (ethylene terephthalate) melts at about 260°C, a rather higher temperature than for nylon 6.6, but the melt does not have the powerful reducing action of nylon melts. The range of stable pigment colorants roughly matches that for nylon 6. Similar methods of coloration using pigment dispersions in the autoclave or for chip coating, as described for nylon fibers, may be used.

The chip-dyeing method practiced with nylon using mainly acid dyes is, however, unsuitable for polyester. Polyester may be dyed with disperse dyes at the boil in the presence of an accelerating agent or carrier or at temperatures of about 130°C, without carrier, under pressure. In the first case, the melt may be contaminated with the carrier; in the second case, melt hydrolysis is liable to occur, because of the difficulty of removing all moisture from the chips after dyeing, neither of which is desirable for the production of good-quality fiber.

Fortunately, an admirable alternative, also giving transparent colored fiber, was discovered by Waters and Wood (British Patent 651,694). The use of predispersed pigments in carriers other than PET is becoming more common (in a process disclosed in a dominating patent which dates back to 1948). They appreciated that the hot molten polymer could, conceivably, be used to advantage as a solvent if colorants could be found which would survive the polymerisation process without sublimation or decomposition, and they devised a test by means of which they proved the point.

Since that time literally hundreds of colorants have failed for either total or partial decomposition in the melt or some particular fastness deficiency, but, also, a number have merited patent protection, and a few colorants are marketed by ICI under the brand name Filomon.

Typical colorants are anthraquinone compounds containing carboxylic acid or hydroxyalkyl groups and phthalocyanine derivatives. Their stability to heat is such that they may be introduced into the autoclave at the monomer-formation stage or they may be added (also in a slurry with a little of the ethylene glycol monomer component) at the end of the ester-interchange reaction, just before polymerisation takes place. The whole process takes several hours at a high temperature. The colorants apparently dissolve readily, but they are, in fact, reactive and copolymerise in the melt.

Despite intensive research, the number of melt-soluble colorants suitable for autoclave coloration has remained small and the color range obtainable rather limited.

Polyester fibers are, however, expensive to dye by the usual high-temperature or carrier methods; there is, therefore a strong incentive to develop mass coloration for them. Work was accordingly continued with a view to increasing the color range and the next step forward was the realization, seemingly by ICI Ltd., in the United Kingdom and by Rhône Poulene Textiles (formerly Société Rhodiaceta) in France, independently, that successful coloration could be achieved if suitable colorants were introduced at the latest possible stage during melt spinning. Reduction in the time of contact of colorant with molten polymer from hours to minutes enables colorants to be used which could not survive the polymerisation process, thus opening up a vast new range of possibilities.

It will be understood that there has been considerable overlap in research and developments in the autoclave, chip-coating, and chip-dyeing methods of mass coloration, but the overall trends of progress have been in the order of which they have been considered. "Late addition coloration" is a distinctive newly developing technology for melt-spun fibers with parallels in the direct injection system for viscose or concentrate blending of plastics, potentially offering similar advantages in simplicity and flexibility of production. The use of pigments is not excluded, but it seems likely that polymer-soluble colorants will be preferred and a promising color range is appearing in the patent literature. For nylon there are, for example, anthraquinone derivatives, and chromium complexes, and for polyester, azo compounds, anthraquinones, and azopyridones.

By the use of melt-soluble colorants with high fastness to temperature and sublimation, all complications caused by coarse pigment particles, e.g., filter blockages, spinneret blockages, filament breakages during drawing, and loss of gloss, are avoided. It is believed, therefore, that products of this kind will establish themselves in the future for the spin coloration of filament and indeed also for polyester staple.

Since the use of the pure colorants is precluded by technical and coloristic difficulties, yarn producers need these melt-soluble products, too, in the form of concentrates in a suitable carrier material. The carrier employed will be either aliphatic or cyclic polyesters soluble in both molten and solidified form in the spinning polymer.

The colorants themselves do not derive from one unified group; they have been selected empirically and are required to have the following properties:

- good resistance to high temperatures (20–30 min. at 280°C)
- good melt solubility (3%)
- good fastness to sublimation
- good textile fastness properties
- good coloristic properties

These may be azo, anthraquinone, polycyclic, or other colorants. For the incorporation of such color concentrates, the points made on the use of pigment concentrates apply.

When pigment concentrates are employed, the far lower temperature stress of 285°C for 20 min. means that many more and cheaper pigments can be used. What is more, the main disadvantage of the condensation method, namely, the enormous cost of cleaning the entire plant, is reduced to that for cleaning the actual spinning equipment. If the sequence of the coloration program is sensibly arranged, cleaning can usually be done by rinsing the equipment through with uncolored material or by following on directly with the next color.

The pigment concentrates can be introduced, for example: by blending the colorless polyester granulate with the concentrate granulate in a tumbler mixer, ideally in the actual storage hopper of the spinning unit; by continuous metering, using conveyor-type weighing machines, straight into the filling inlet of the spinning extruder (granulate to granulate), by addition in molten form via side-mounted screw extruders into the melt zone of the spinning extruder (melt in melt); and by addition in molten form via metering pumps (e.g., gear pumps) into the flow path of the spinning melt in the case of continuous-running condensation spinning units or the older coiled-grid spinning units, this being followed by statically or mechanically operating mixers.

Although there has been great progress compared with the condensation process described previously, substantial difficulties remain to be overcome. Polyester is a highly sensitive material which reacts very vigorously with extraneous substances, leading to depolymerization and consequent changes in viscosity, which in turn cause considerable spinning difficulties. For this reason, the spinning polymer itself would be the most suitable carrier material for pigments from the fiber producer's viewpoint, though not from the viewpoint of the concentrate manufacturer. Spinning polyester has a high melting range, high melt viscosity, high sensitivity to moisture, low wetting capacity vis-a-vis organic pigments, and low pigment uptake. It is thus highly unsuitable as a carrier material.

Hence, besides overcoming the pure dispersion problems, a suitable carrier must be found for development of a concentrate range. Besides being compatible with the spinning polyester, this carrier must be one in which pigments can be easily and effectively dispersed in concentrations up to 40–60%. A few companies have already solved the problem and have corresponding concentrates on offer.

Another method of spin-coloring polyester is provided by fluorescent colorants; these differ from the melt-soluble colorants just described not in their processing technology but in their effect on the materials they are used to color.

These products are pure colorants which dissolve well in various plastics, including polyesters, to give very high-visibility, fluorescent colorations with excellent fastness properties, especially fastness to light. The fluorescent colorants can be combined with one another, as well as with selected pigments or melt-soluble colorants, so that virtually all interesting shades can be obtained, especially those for the safety sector.

When fluorescent colorants are employed, it should be borne in mind that for any particular yarn count and filament thickness the optimum fluorescence can be attained only in a particular colorant concentration range. In other words, an increase in fluorescence is possible only so long as the total quantity of light suitable for stimulating fluorescence is utilized (fluorescence optimum). If the concentration is raised above this optimum the fluorescence declines markedly. This reduction in fluorescence is due, amongst other things, to the increasing individual absorption of the colorant molecules, since the absorption range and emission range overlap. In some instances, overdosing is accompanied by a decrease in light fastness and fastness to heat setting.

Fluorescent colorants can alternatively be ordered in the form of color concentrates. Their main possibilities of use appear to us to be in the safety sector.

Woven and knitted fabrics made from polyester materials containing fluorescent colorants would considerably increase the visible warning effect of children's clothing in road traffic, working clothes on the road and railway, life jackets, life rafts, etc.

POLYOLEFINS

Polyolefin fibers are fibers spun from polymers or copolymers of ethylene and propylene. Their origin was the successful polymerization of ethylene in the early 1930s, but, although the use of polyethylene later expanded from plastics into textile fibers, this polymer has been eclipsed for textile usage by polypropylene. The first polypropylene fibers were manufactured by Montecatini in Italy, in 1957.

Polypropylene is, practically speaking, the only polyolefin fiber on the market now. The amount of polypropylene produced, worldwide, for textile purposes in staple-fiber, continuous, monofilament and multifilament yarns, tape, and fibrillated film forms is estimated to have been not less than 2 billion lb. Most of the colored fiber used, whether melt spun or from fibrillated film, is mass pigmented at the melt stage.

The reason why a large proportion of this tonnage would, undoubtedly, be mass colored is because the homopolymer of propylene is undyeable. Also, the various dyeable types, which have become available, are reported to be making little headway because pigmented fibers are well established and preferred for their superior light fastness and the simpler fabric-finishing methods.

Since pigments are used exclusively for mass coloration, equipment for their incorporation is an integral part of most production units. The choice of pigment is governed to some extent by the form of the product, whether fine filament (down to 3.3 decitex), or tape, or film (typically of 50- to 80-μ thickness).

For fine filaments the molten polymer is extruded at temperatures as high as 300°C, although its melting point is about 165°C, and consequently the pigments used must have high thermal stability. They are, therefore, similar to those used in nylon and polyester. As with these melt-spun fibers the pigments may be incorporated in the bulk polymer at spinning concentration, or highly pigmented masterbatch chips or granules may be prepared for subsequent blending with uncolored polymer before extrusion.

The extrusion of film for slitting into ribbons for tape yarns or for fibrillating is, in its essential details, typcial of techniques used in the plastics industry and a considerable proportion of the total production is, in fact, manufactured by plastics firms that have diversified their output by installing equipment, including drawing machinery, for these textile applications.

For extrusion through flat dies (slits) the molten polymer is heated to 200°-240°C and so heat-stability requirements are somewhat less exacting than for the production of fine filaments. Particle-size requirements are, also, rather less stringent because of the larger cross-sectional area of the orifices of the dies and of the finished materials. Pigment dispersibility is, however, particularly important because high-speed, large-capacity machinery is normally used. For this purpose specially predispersed pigments have been developed, consisting of extremely fine pigment particles coated with, or trapped in, a compatible resin such as polyethylene, and their processes of manufacture have been patented. Ranges of these products are available from colorant manufacurers, for example, for ICI, BASF, Ciba-Geigy, Hercules Inc., Ampacet, Inmont, etc. Guidance on the choice of colorant, application procedure, and fastness properties is available from these companies. Predispersed colorants eliminate the need to grind the pigment into the polymer to develop maximum strength and brightness. They enable fine dispersions to be produced quickly using simple equipment, and they alleviate dusting problems which can occur with the parent pigments.

Because of the lower melt temperature, a wider range of pigment types (not all necessarily available in predispersed form) is suitable,

Table 17. Inorganic pigments suitable for polypropylene.

White	*Yellows and Orange*
Titanium dioxide, rutile	Cadmium sulfo-selenide
Zinc oxide	Chrome yellow
Antimony oxide	Chrome orange
Black	Molybdate orange
	Strontium yellow
Channel black	Zinc chromates
Furnace black	Ni-titanium yellow
Violet, Maroons, and Reds	Iron oxide
	Neaple yellow
Mercadium reds	Primrose cadmium yellow
Cadmium selenide	Mercadium yellow, orange
Strontium rubine red	Zinc titanate yellow
Iron oxide	(Zinc-iron) and titanium antimony yellows

Notes:

1. Titanium dioxide pigments should be used in low concentrations for delustering only. Zinc oxide can be used at high concentration. Antimony oxide comes in question mainly for flame retardant compositions.
2. Furnace blacks have generally replaced channel black in the United States. Beware of the quality of dispersion!
3. Cadmium and mercadium yellows and oranges are excellent at all pigment levels but are not recommended for outdoor uses.
4. Lead chromate yellows and molybdate oranges have poor heat stability limiting their use to low extrusion temperatures.
5. Strontium rubines are limited to deep shades and low extrusion temperatures. However, the color is inexpensive and should not be overlooked for special applications. Iron oxide can be used for rust and gold colors but care should be taken with regard to stabilization.

Table 18. Organic pigments suitable for coloration of polypropylene.

Violet, Maroons, and Reds	*Orange and Yellow*
Quinacridone red	Benzidine yellow analine
Quinacridone violet	Benzidine yellow HR
B. O. N. red (2B. Ba salt)	Benzidine orange
B. O. N. red (2B. Ca salt)	Dis-azo yellow, orange, red, brown
Helio bordeau maroon	Pyranthrone orange
Thioindigo red	GR parinone
Thioindigo violet	Isoindilinone yellow, orange
Dioxazine violet	Flavanthrone yellow
Red lake R	Anthrapyrimidine yellow
Anthraquinone red	
Isoindolinone red	*Blue and Green*
Pyrylene reds	Phthalocyanine blues
Isoviolanthrone violet	Phthalocyanine greens
Perinone red	Indanthrone blue

Notes:

1. Perylene reds and dis-azo reds are excellent in deep shades. Perylene reds are excellent in most concentrations for tints and beiges.
2. Organic yellows are best for deep, high-luster colors but should be avoided for tints and beiges.
3. Quinacridone and carbazole violet pigments have excellent pigmenting properties in full shades and in tints.
4. Phthalocyanine blues and greens are excellent for all turquoise and aqua, and for full shades or tints of blue.

including chrome pigments, toners, and inexpensive azo pigments such as C.I. Pigment Red 5 and C.I. Pigment Red 7, in addition to the ubiquitous phthalocyanines, but all must be carefully selected to avoid the defect of migration on prolonged storage, as well as to meet the more usual fastness requirements.

Table 17 and Table 18 list some of the inorganic and organic pigments, respectively, found suitable for mass coloration of polypropylene.

CHOICE OF PIGMENTS AND COLOR MATCHING

In selecting a specific pigment for use as a colorant for synthetic fiber, consideration has to be given to many factors pertinent to processing conditions and end-use environments of the product. A selection based solely on hue, brightness, tinctorial strength, transparency, etc., would in many cases lead to disasterous affects. Thus, once the tinctorial properties of a pigment have been established, consideration must be given to its fastness properties, its resistance to heat, light, moisture, household chemicals, etc., and its interactions with the polymer or with the stabilization system. These requirements are briefly discussed in this section with reference to polypropylene as an example.

Color Characteristics

The color characteristics of a pigment include a description of its opacity, relative tinctorial strength, and optical effects (e.g., irridescence and metamerism), in addition to its spectrophotometric response. Tinting or tinctorial strength of a pigment is defined as the degree to which the colorant imparts color to a standard white pigment. It corresponds to the color yield of a pigment. The manufacturers of pigments generally furnish information on tinctorial strength of pigments in the form of their tinting strength ratings. These are obtained by grinding a weighed amount of colorant into a lithographic varnish for a specified length of time. This procedure, however, provides no tangible information on the dispersing properties of the pigment in polypropylene and, therefore, is not a reliable index of tinting strength obtained in the coloration of polypropylene fibers.

Although the degree of dispersion has a profound effect on the tinting strength and hue of the pigment, the primary basis for determining its tinctorial characteristics is the chemical and crystalline structure of the colorant. Some pigments are inherently weak in tinctorial strength; ultramarine blue, cadmium sulfoselenide, cadmium mercury pigments, iron oxides, molybdates, cobalt aluminate, barium, strontium, and zinc chromates, etc., can be cited as examples.

Closely related to tinctorial strength is the hiding power of a pigment. This property is of primary importance in the paint industry; in coloration of fibers, the hiding power is important mainly in the case of white pigments which are used as delustering agents. The hiding, covering, or delustering power of these pigments is a function of their refractive index.

Weatherability

Weatherability refers to the ability of a pigment to withstand all forms of weathering, including visible and UV light, moisture, temperature, and chemicals in whatever combination they may exist during the life of the product. Weatherability traits of importance include color fastness, resistance to chalking, and retention of physical properties. Weatherability is often a function of the light stability and oxidation resistance of a pigment since these elements are encountered in the course of weathering.

No matter how good its other properties, a pigment is of little value if it will not withstand the effects of exposure to light and weathering. This is specially true for fibers, because the very small cross section of the filament makes the pigment more vulnerable to the action of light and other weathering agents than the thick cross-sectional forms generally encountered in the plastic industry. A large proportion of polypropylene fiber is currently being used in home furnishing and outdoor uses which require particularly prolonged exposures to weathering conditions. In fact, it was the exceptional light fastness of polypropylene products which was offered to the consumer as one of the primary strong points of polypropylene over the other manmade fibers to gain a sizeable proportion of the home furnishing market.

The light stability ratings of colorants are often expressed as a 1 to 8 scale: 1 being the poorest and 8 being the highest. Most of the pigments offered commercially are known to command high-lightfastness ratings. The quoted ratings for lightfastness are, however, not wholly reliable, as light stability of most of these pigments in polypropylene does not correspond with the relative ratings quoted by the pigment manufactureres in other polymeric systems. This is because of a strong interaction of many pigments with a polypropylene substrate, the presence of various stabilizers and antistatic additives, and the obviously thin cross section of the fiber. In general the lightfastness ratings of pigments are lower in polypropylene than those in other substrates. A careful testing is, therefore, needed before a pigment can be recommended for production use. Inorganic pigments generally have excellent stability to light. The organic pigments range in quality from poor to excellent according to their structure. Furthermore, a pigment of poor stability to light will appear to be affected more significantly in a formulation containing a white pigment, such as TiO_2.

Fading may be due either to a loss of color of the pigments, to fiber becoming white through surface crazing, or to fading or discoloring of stabilizers present in the fiber. The color change due to fading or discoloration of stabilizers is important only in the case of some nickel containing stabilizers and to some extent in the case of simple phenolic antioxidants. The problem is not serious in most of the cases.

Some pigments show little change in color on exposure to light alone, but they fail rapidly when exposed to outdoor conditions. Cadmium-based pigments can be cited as examples. When used in interior applications they exhibit excellent stability against weathering, both in mass tone and tint applications, to both sunlight and artificial light. Outdoors, in tints or pastels, fading occurs rather rapidly. These changes (fading, darkening, or change in hue) are usually due to the presence of reactive groups on the pigment which are affected by moisture and atmospheric chemicals.

Light fading is often dependent upon the concentration of the pigment employed. As is true in the case of colored fibers of other origins, pastel shades exhibit lower fastness ratings than the heavy shades. The degree of dispersion has also a profound effect on the light stability of the pigment. Highly dispersed pigment would fade more rapidly than poorly dispersed pigment. The inclusion of UV absorbers in pigment formulation to improve lightfastness usually gives disappointing results, probably because the amount of absorber in the thin layer of polymer between the surface of the fiber and the first pigment particles is too low to be effective.

Heat Stability

Heat stability is the resistance to change in color or performance as a result of heat encountered during processing. These changes may result from chemical decomposition, crystal change, or interaction with other ingredients in the compounded polymer. Heat stability is a function of temperature and time and may affect performance without affecting color.

The temperature and time of compounding during melt spinning of pigmented polypropylene resin into fibers may vary from a few minutes at 250°C to 30 min. at 300°C. Moreover, the environment is seldom chemically inert; acid, oxidative, or reductive conditions may exist or develop owing to the thermal breakdown of the system. The combined effect of time, temperature, and reactive medium frequently leads to marked color changes in the coloring matter, or even to its complete breakdown. The degree to which a colorant withstands these conditions is termed its stability to processing or more simply, but less accurately, its heat stability.

Because of the high processing temperatures employed in melt spinning, excellent resistance to heat is a prerequisite for a pigment to make it useful for coloration of polypropylene fibers. Most organic pigments can withstand a few minutes at high temperatures, in a range of 250° to 300°C, although prolonged exposure leads to decomposition, darkening, or sublimation. In contrast to the generally poor heat resistance of organic pigments, inorganic pigments exhibit excellent resistance to heat. High spinning temperatures eliminate most organic pigments and the pallete of useful pigments for polypropylene, therefore, comprises of mostly inorganic pigments. Of the useful organic pigments stable at the processing temperatures of polypropylene, phthalocyanine green and blue and carbazole violet possess the greatest overall utility, followed by quinacridones, isoindolinones, and perinones. The greatest deficiency is noted in the red and yellow regions of organic pigments. This is certainly an area for future innovative organic pigment technology.

Chemical Resistance

Chemical resistance refers to inertness and compatability with other ingredients present within the compounded polymer as well as resistance to the external environment. It is often associated with heat stability since reaction may take place during high-temperature processing that would not otherwise occur.

Some of the commonly used pigments or the impurities present in them may react with such other additives as stabilizers, antistatic agents, and processing aids present in the compounded polypropylene. The reaction of sulfur-bearing pigments or impurities with nickel stabilizers and the reaction of lead impurities with sulfur-bearing stabilizers are well known. These reactions normally result in severe discoloration of the melt or, in minor cases, in a dull shade. Chrome pigments are normally contaminated with a lead compound. They cannot be used, therefore, in combination with those stabilizers which may produce colored sulfides as a result of their interaction. The same is the case with neaple yellow (lead antimonate). Cadmium reds and oranges, as well as mercadium reds, are not suitable in mass tones in conjunction with nickel-complex stabilizers, since the formers are based on sulfides.

There have been reports of chemical interaction of certain pigments with a few antioxidants giving rise to yellow-colored products. For example, it has been shown by extraction and chromatographic studies that a portion of 4,4-methylene-bis(2,6-di-t-butylphenol) was converted into the hydrogalvinoxyl structure when repeatedly brought in contact with channel black in a refluxing hexane solution. It is believed that other phenolic antioxidants are oxidized to quinoid-type structures in the presence of certain pigments. Ultramarines are sensitive to hydrochloric acid. If the concentration of polymerization catalyst residue to significant in the polymer, the HC1 liberated from $TiCl_3$ of the catalyst residue may interact with the pigment to change its color, hue, or intensity.

Apart from the nickel-complex stabilizers, sulfur-bearing compounds, and catalyst residues, there is also the possibility of interaction of pigments with some other additives or even with other pigments. Common antistatic agents used in polypropylene formulation, for example, behave as reducing agents at high temperature, discoloring certain organic pigments that are normally stable at the processing temperatures. Some stabilized polypropylenes will "base-exchange" with alkaline earth metal toners, changing their color from bluish red to orange. Metal free phthalocyanines will complex with metals (normally lead) from other pigments present in the formulation.

Catalytic Action on Polypropylene Stability

In addition to thermal and photodecomposition of colorants to colorless or off-color compounds, some of the colorants are in a position to sensitize the decomposition of polypropylene under the action of light and heat. There are also some colorants which act favorably in this respect and retard the decomposition of the polymer. Inorganic pigments are particularly noticeable for their catalytic action on polypropylene stability, while some organic pigments are also known to accelerate the decomposition of polypropylene by supplying free radicals to the polymer decomposition chain. Some pigments, particularly carbon black and zinc oxide, protect the fiber from photodecomposition by virtue of their high degree of screening power; other organic pigments may help to retard the oxidative degradation of polypropylene by virtue of their ability to scavenge the free radicals produced in the process of oxidative decomposition. Careful attention is, therefore, necessary in the selection of pigments with regard to their degradative and stabilizing action of the polymer during processing and during the useful life of the product.

The effect of pigments on thermal and photodegradation of unstabilized polypropylene has been extensively studied by several authors. Uzelmeier has conducted detailed studies on the effect of pigments on the heat and light stability of stabilized polypropylene. Table 19 summarizes the changes in heat stability brought about by various types and amounts of some common pigments relative to an unpigmented stabilized control sample. The data show that none of the pigments improve the heat stability of polypropylene in the presence of an effective phenolic antioxidant. In fact, serious deleterious effects are noticed in several cases. The pigments, listed in order of decreasing antagonism in Table 19 may be roughly divided into three classes – those that are severely prodegradant, including channel black, chromophthal red 3B, quinacridone maganta, phthalocyanine blue, and iron oxide tan; those that are moderately prodegradant, including phthalocyanine green, mineral and furnace blacks, ultramarine blue, and chromium oxide; and those that are slightly prodegradant, including cadmium yellow, mercadium red, and rutile titanium oxide. Of special interest is the channel black which acts as a mild antioxidant in the absence of phenolic antioxidant but causes a 50% reduction in heat stability of stabilized polypropylene with as little as 0.10% addition.

Uzelmeier determined the relative rates of hydroperoxide decomposition for various pigments and compared them with their relative prodegradant activity in stabilized polypropylene. No correlation was found. It thus appears that the reduced stability observed with

Table 19. The effect of pigments on the heat stability of stabilized polypropylene.

Pigment	phr	Decrease in Heat Stability (%)[a]	Pigment	phr	Decrease in Heat Stability (%)
Channel black	.01	17	Mineral black	.10	46
Channel black	.05	37	Mineral black	.50	25
Channel black	.10	56	Mineral black	2.0	49
Channel black	.25	76	Furnace black	.10	15
Channel black	.50	93	Furnace black	.50	29
Channel black	1.0	95	Furnace black	2.0	51
Cromophtal red 3B	.10	29	Ultramarine blue	.10	15
Cromophtal red 3B	.50	78	Ultramarine blue	.50	15
Cromophtal red 3B	2.0	95	Ultramarine blue	2.0	49
Quinacridone magneta	.10	29	Cr_2O_3 green	.10	5
Quinacridone magneta	.50	51	Cr_2O_3 green	.50	17
Quinacridone magneta	2.0	85	Cr_2O_3 green	2.0	49
Phthalocyanine blue	.10	22	Cadmium yellow	.10	12
Phthalocyanine blue	.50	54	Cadmium yellow	.50	27
Phthalocyanine blue	2.0	85	Cadmium yellow	2.0	10
Fe_2O_3 tan	.10	39	Mercadium red	.10	15
Fe_2O_3 tan	.50	63	Mercadium red	.50	15
Fe_2O_3 tan	2.0	85	Mercadium red	2.0	15
Phthalocyanine green	.10	29	Rutile TiO_2	.10	10
Phthalocyanine green	.50	44	Rutile TiO_2	.50	15
Phthalocyanine green	2.0	56	Rutile TiO_2	2.0	12

[a]Relative to unpigmented, heat-stabilized control; 10-mil films, days to embrittlement in a circulating air oven at 150°C

Uzelmeier, C. W., SPE Conf. on "Pigmentation," New York, (1970).

Table 20. The effect of pigments on the light stability of stabilized polypropylene.

Pigment	%	Increase In UV Stability (%)[a]	Pigment	%	Increase In UV Stability (%)[a]
Channel black	.01	70	Rutile TiO_2	.10	60
Channel black	.05	220	Rutile TiO_2	.50	290
Channel black	.10	370	Rutile TiO_2	2.0	600
Channel black	.25	890	Fe_2O_3 tan	.10	30
Channel black	.50	900	Fe_2O_3 tan	.50	160
Channel black	1.0	1300	Fe_2O_3 tan	2.0	610
Phthalocyanine blue	.10	170	Cr_2O_3 green	.10	40
Phthalocyanine blue	.50	600	Cr_2O_3 green	.50	140
Phthalocyanine blue	2.0	900	Cr_2O_3 green	2.0	510
Mercadium red	.10	70	Quinacridone magenta	.10	30
Mercadium red	.50	390	Quinacridone magenta	.50	80
Mercadium red	2.0	890	Quinacridone magenta	2.0	160
Cadmium yellow	.10	70	Cromophtal red 3B	.10	20
Cadmium yellow	.50	380	Cromophtal red 3B	.50	80
Cadmium yellow	2.0	890	Cromophtal red 3B	2.0	160
Phthalocyanine green	.10	80	Mineral black	.10	50
Phthalocyanine green	.50	290	Mineral black	.50	40
Phthalocyanine green	2.0	900	Mineral black	2.0	70
Furnace black	.10	20	Ultramarine blue	.10	0
Furnace black	.50	190	Ultramarine blue	.50	0
Furnace black	2.0	890	Ultramarine blue	2.0	0

[a]Relative to an unpigmented, stabilized control; 10-mil films, Carbon Arc Weather-O meter hours to brittle flex failure.

Uzelmeier, C. W., SPE Conf. on "Pigmentation," New York, 1970.

many pigments in stabilized polypropylene is caused by interactions of pigments and stabilizers rather than the peroxide decomposition or chain starting power of the former. One type of interaction that has been considered is the absorption of stabilizers by pigments.

Because of their light-screening power, pigments are expected to contribute favorably toward light stabilization of polypropylene. Table 20 shows the effect of pigments on the light stability of stabilized polypropylene. It is shown that the light-protective action of pigments is generally realized at a high level of pigment loading but it is generally negligible in thin cross sections or at a low level of pigment loading. It has been shown that some pigments, instead of protecting the polymer from actinic degradation, accelerate its decomposition.

It follows from the above discussion that pigments should be carefully examined for their interference with the stability of the polymer. The use of those pigments which excessively accelerate the heat and light degradation of the polymer should be avoided. When a catalysing pigment has to be used for economic or tinctorial reasons, a better degree of stabilization would be in order. It follows further that, to meet the quality control standards of thermal and photostabilization of the fiber equally with each color, the stabilization formulation must be adjusted to take into account the deleterious effects of the pigment formulation on polymer stability. This adjustment is recommended not only when the nature of the pigment changes in the formulation but also when the concentration of the same pigment changes to a considerable degree.

Effect on Tensile Properties

Pigments can have significant effect on the mechanical properties of the fiber. Most of the pigments tend to impair the drawability and tensile properties of polypropylene fibers. Similarly it has been shown that increasing amounts of pigments increasingly impair tensile properties. The deleterious effect of pigments on tensile properties of the fiber may have several reasons. The pigment particles and agglomerates can serve as weak points in the structure, the physical properties can be impaired through the induced decomposition of the polymer in the presence of pigment, or the fiber properties may be affected through morphological changes in the fiber. In the latter case, the drawability may not always decrease and the tensile strength may even experience a slight gain.

Rheological Properties

Flow properties of polypropylene melt are sometimes profoundly affected by the presence of certain pigments in the formulation. This is mainly due to the stabilizer pigment interaction or to catalytic decomposition of the polymer, as described previously. The change of rheological properties through internal lubricating action of pigments is not known. Extensional flow properties of polypropylene are generally affected more by the presence of pigments than by the shear flow properties. An enhanced "rain effect" (nonuniform draw-down of the molten filament) is a common phenomenon. The cause of this kind of flow disturbance is not fully understood but it is theorized that the uneven drawability of the molten jet may be due to microscopic inhomogeneity in polymer viscosity. These microscopic inhomogeneities in polymer viscosity may be caused by localized catalytic degradation of the polypropylene surrounding the functional pigment particle. Another obvious possibility is the poor compatibility of the pigment or the pigment carrier with the polymer.

Morphological Properties

Some pigments, by virtue of their nucleating power, may influence the morphological structure of the polypropylene fiber. The fibers, composed of a large number of relatively small crystallites, are expected to show pronounced shrinkage in boiling water, and the tape yarns of such morphological structure should have a greater tendency of fibrillation. Phthalocyanine blues and greens are known to produce these effects. On the other hand, quinacridone reds are known to considerably increase the rigidity and decrease the impact strength of polypropylene particles.

Resistance to Migration

There are three types of colorant migration which are important for polypropylene fiber: 1) bleeding: the migration of the colorant from the fiber to the material with which it is in contact. Poor wash and dry cleaning fastness is the obvious result of bleeding. Pigments generally do not suffer from this defect. 2) crocking: migration of the colorant to the surface of the fiber so that it can be rubbed off. Crocking may be caused by pigment migration, but may also result from poor dispersion, high pigment loading, or abrasion or deterioration of the fiber surface, such as through weathering. Poor rubbing fastness, either dry or wet, is the result of crocking. This property is specially important in the case of polypropylene fiber because quite a substantial portion of polypropylene fiber is being used for carpeting and upholstery fabrics. Inorganic pigments are generally not of much concern in this respect; however, some organic pigments may lack an adequate crock fastness. 3) Blooming, chalking, bronzing: a chalky or metallic effect that is produced when a pigment migrates to the surface of the fiber, either in high temperature processing (bronzing) or cold drawing (chalking). This phenomenon is almost specifically associated with organic pigments which are not compatible with polypropylene because of their hydrophylic groups. The blooming of pigments is of special concern in those uses where the produce is going to be subjected to some sort of heat treatment, such as autoclaving and latexing. During cold drawing, pigments may bloom or chalk and be deposited on draw rolls and yarn guides.

Dispersibility

Pigment manufacturing processes are generally designed to produce a particle size that will yield optimum spectral characteristics. This so-called primary particle size ranges from a few hundredths of a micron for most organic pigments to 1 μ or more for some inorganics. However, for various reasons, these primary particles tend to aggregate or agglomerate during manufacture to form clusters of particles several times their intended size. The pigment powder thus consists of aggregates ranging in size from a few microns up to perhaps 200 μ. If the user is to obtain full value from the pigment in terms of strength development, transparency, depth of color, brightness, and opacity, sufficient work must be done on the pigment-binder system to disperse these agglomerates to something approaching the particle size of single crystals formed in the original precipitation or calcination process.

Dispersibility describes the ease with which pigment aggregates can be reduced to an acceptable particle size. The process of dispersion involves the production of discrete particles, as near as practical to their primary particle size, and maintaining them in that condition. The practical aspects lie in recognizing that the dispersion process is often costly, and that reducing the pigment agglomerates to their absolute primary particle size is not essential or possible in practice. The optimum particle size for each pigment will vary, and pigments succeptible to overgrind are known. In most cases the optimum particle size of pigment dispersion for good optical properties generally lies in the range of 0.3 to 0.8 μ.

To achieve even coloration, two criteria must be satisfied: the color must be evenly distributed throughout the medium and the colorant must be adequately dispersed. Dispersion in this context refers to methods for wetting out and reducing in size the agglomerates and aggregates of dry pigments and not to means of uniformly distributing it in the system after dispersion. It is normally not difficult to control even distribution but problems often arise in dispersion.

Colorists and pigment technologists are aware that the degree of pigment dispersion in a medium has a direct effect on visual properties. Transparency increases with reduction in aggregate size and approaches a maximum as the degree of dispersion reaches the optimum. Tinctorial strength, and along with it the economy of shade, is affected in the same fashion. What is perhaps not as fully appreciated is the fact that changes in hue also are directly related to dispersion. Blues become cleaner and greener as the dispersion improves, greens and reds become cleaner and yellower. Obviously, in any situation involving close color control from lot to lot, it is of paramount importance that the pigment dispersion be of consistent quality.

In addition to the problems of color intensity and hue, the degree of dispersion has a pronounced effect on the filtration characteristics of the melt. A poorly dispersed pigment would lead to rapid clogging of the filter pack and necessitate frequent interruptions in spinning for filter change. Additional direct effects of inadequate dispersion include high cost, increased color control problems, poor mechanical properties of the fiber, tendency to chalking and crocking, and increased number of thick-and-thin filament sections.

Pigments differ in their ease of dispersion. Pigments such as titanium dioxide and cadmium or chrome pigments normally are among the easiest to disperse; others, such as carbon blacks and iron oxides, are the most difficult. Practically any pigment which has a particle size of less than about 0.2 μ will show appreciable aggregation during processing and will present some problem in dispersion. The severity of aggregation is also dependent on the type of pigment. Hydrophilic pigments aggregate severely when dried from aqueous media; organophilic pigments have less tendency toward aggregation as the particle size is reduced.

Filtration Characteristics

Excessive pressure rise in melt filtration during the spinning of pigmented polypropylene fibers is an ever-present technical problem, necessitating a frequent change of filter and causing a large amount of wastes, high labor costs, and increased periods of equipment shutdowns. Filtration characteristics of a pigmented polypropylene blend is a function of the degree of pigment dispersion and, probably more importantly, is dependent on the uniformity of dispersion. A material with a few, very large agglomerates may not perform in practice as well as one containing a large number of smaller agglomerates. All those factors which influence the degree and uniformity of pigment dispersion also affect the filtration characteristics of the blend.

Knife Blunting

In the production of polypropylene staple fibers the hardness of the pigment particles results in greatly increased wear on the cutter's knives, resulting in blunting of cutting roller edges in pacific converters. This gives rise to fused and hooked staple ends, causing difficulties in staple processing into spun yarns. Similarly, in tape yarn processes the hardness of pigments is reflected in the increased wear of film slitting blades and fibrillating pins. The effect of blunt slitting blades is seen as ragged edges on the tapes which cause tape snaps during orientation or weaving.

Pigments differ greatly in their knife-blunting effect, depending on their hardness and the quality of dispersion. Some pigment formulations would give a knife life of several days while the others would necessitate a blade change within a few hours. Knife life is, therefore, similar to filter pack life. All pigments increase knife blunting, but organic pigments are less damaging than the inorganics. Among inorganic pigments, titanium dioxide, iron oxides, and chromium oxides are the hardest and the most damaging with respect to knife wear. Carbon black, cadmium sulfide, chrome yellows, and zinc sulfide are less damaging.

Cost Effectiveness

One of the most important considerations, always kept in view in color formulation, is the economy of coloration. Sometimes alternatives do not exist because of the particular demands on processing or tinctorial and color fastness properties of the formulation. But most often there is the alternative of choosing one out of two or more possible pigment candidates for a specific end-use and particular, although not the best, processing and color fastness properties. The overall efforts in these cases is to minimize the cost of coloration while meeting the minimum color fastness and processing specifications. Pigments differ widely in their prices. The price tag is influenced by two main factors: 1) the chemical nature of the colorant, and 2) the physical state of the colorant, e.i., whether it is sold as dry color, treated dry color, resin dispersion, color concentrate, solvent dispersion, or aqueous dispersion.

A distinction should be made between the price of the pigment and the price of coloration. A low-price pigment may not necessarily result in a low price of coloration. For example, although the dispersions are comparatively more expensive than the corresponding dry colors, the use of the former may end up in a low pigment concentration needed for a comparable shade, because of the higher color yield of well-dispersed colorants. Very often the higher price paid for a well-dispersed pigment can be offset by improved spinning continuity and low waste. Similarly, the color concentrates may appear expensive in comparison with color dispersions or dry colors, but the savings in blending procedures may offset the additional cost paid for color concentrates.

There is still no hard and fast rule to determine whether inorganic or organic pigments should be used if offered the choice. The former are relatively inexpensive and have excellent resistance to heat, light, and migration. However, they have lower tinctorial strength than the organics. At present, almost all of the modern high-performance organic pigments are rather expensive and have not automatically replaced inorganics in comparative formulations. But by virtue of their high-tinctorial power, organic pigments may be cheaper in the ultimate analysis for some cases.

TESTING AND EVALUATION

The evaluation of a pigment formulation can be grouped into three classes of tests:

1. Process evaluation (heat stability, filter plugging rate, change of color with processing temperature, interaction with stabilizers and other additives, catalytic degradation of polymer, spin continuity, sublimation, chalking on drawing, etc.).
2. Weathering evaluation (effect on light stability of polymer formulation, change in color in polluted atmosphere, etc.).
3. Common textile testing (dry cleaning, wash fastness, light fastness, gas fading, crocking, bleading, etc.).

The actual details of process evaluation procedures practiced by various fiber manufacturers differ, and it is largely proprietory information. The essence of all these evaluation procedures is the actual production of the fiber and yarn on laboratory or pilot plant equipment under varied processing conditions, from blending to extrusion to texturizing or staple cutting, then observation of the various steps for any anticipated shortcomings in the pigment formulation. A 1½-in. diameter extruder, fitted with a convenient filter pack and spinnerette assembly and equipped with spin-draw arrangement, is a convenient tool for routine testing for process evaluation. For heat stability evaluation, a series of melt temperatures and throughputs are employed and the subtle changes in color of the fiber and drastic changes in melt-flow index or intrinsic viscosity of the extrudate noted. The same applies to sublimation and chalking resistance, although it may take long spinning runs to detect these deficiencies. The filtration characteristics are normally determined by measuring the rate of pressure rise in the spinnerette pack using a small area of multiple layers of progressively finer screens (up to B. S. N. 400 mesh). The base polymer (stabilized) is taken as the comparison point.

Measuring and counting pigment particles in the extrudate by such techniques as the Quantiment particle-size analyzer gives a good approximation of performance, but it suffers from the fact that the sample examined is very small. The number of agglomerates causing spinning troubles in a reasonably good material is generally so small that they can easily be missed in a small sample. Break index (counts of threadline breaks per 100 lb of yarn) or minor break index (the counts of filament breaks without causing a break in threadline) is another indication of the degree of dispersion and compatibility of the pigment with the polymer. A change in tensile factor ($TE^{1/2}$) caused by addition of a pigment can also be used as an indication of the degree of dispersion, compatibility, effect on polymer stability, rain-effect, and morphological effects of a pigment.

The weathering tests for the evaluation of pigments for their effect on light stability are common with those employed for the testing of stabilizers and are generally undertaken concurrently with the latter. In these tests, the changes in color as well as tensile properties are observed. Some pigments are very stable to the action of light alone but they are poor under the simultaneous action of light and moisture. Similarly, some pigments are sensitive to sulfur pollution of the atmosphere. This makes the free weathering test mandatory. The common textile tests for evaluation of lightfastness, gas fading, etc., are well known and are listed in Table 21.

Table 21. Common textile test methods for pigment evaluation.

Test	AATCC Test Method
Acid and alkali fastness	6–1957
Bleaching with chlorine	3–1962
Crocking fastness	8–1968
	116–1966
Dry cleaning fastness	85–1968
Light fastness	16–1964
Phototropism	32–1952
Gas fading	75–1956
Perspiration fastness	15–1962
Wash fastness	61–1968
Water and lightfastness	125–1967

BIBLIOGRAPHY

1. Sunn, G. F. "Coloration of Polypropylene Fiber and Film." SPE Annual Technical Conference (1969).
2. Bash, D. P. "Spin Coloration of Polyester and Polyamide Fibers." SPE Annual Technical Conference (1975).
3. Ackroyd, P. "The Mass Coloration of Man-Made Fibers." *Rev. Prog. Color. Relat Top.* **5**, 86 (1975).
4. Nevls, R. W. "A New Era for Nylon Fiber Mass Coloration with Pigments." SPE Annual Technical Conference (1975).
5. Ahmed, M. *Polypropylene Fibers – Science and Technology,* Chapter on "Pigmentation." Elsevier, Amsterdam, Holland (1979).

Glossary

APPEARANCE. The way things look. Color is an aspect of appearance.

BEER'S LAW. The absorption of light at each wavelength is proportional to the amount of the absorbing material.

BLEEDING. The migration of colorants or other additives from one plastic to another or to a solvent.

BLOOMING. The migration of colorants and other additives from within a plastic to a free surface.

BRONZES. Copper-based metallic flake pigments.

BRONZING. Colored reflection.

CHROMOPHORES. Color-producing chemical groups in the molecule of a pigment or dye.

CIE. Commission International de l'Eclairage.

COLOR. Color is the brain's perception of light that has been modified by the absorption of part of its rays.

COLORANT. A substance that modifies the perceived color of an object, or imparts color to an otherwise colorless object. Colorants are subdivided into *dyes* and *pigments*.

COLORED CONCENTRATES. Concentrated dispersion of a pigment or a dye in plastics.

COLOR COORDINATES. Hue (dominant color), value (lightness vs darkness), and chroma (saturation or purity of color). The observed differences in color between two objects can be described in terms of these elements.

COLOR YIELD. See *Tinctorial Strength.*

COVERING POWER. The ability of a pigment to opacify an object.

DAYLIGHT FLUORESCENT DYES. See *Fluorescent Dyes.*

DELUSTERING AGENTS. Opacifiers.

DICHROISM. Change of hue with a change of concentration of colorant in the colored material.

DISPERSION. An intimate mixture of finely dispersed pigment in a carrier. The carrier can be a resin, a liquid, or a wax.

DRY COLORING. Coloring of plastics using dry pigments.

DYES. Synthetic organic colorants, soluble in water or other solvents including polymers.

EXUDATION. Migration of additives from the interior to the surface of the plastic. Exudation can occur at room temperature and is generally confined to liquid additives.

FLUSHED PIGMENTS (treated pigments). Pigments treated with surface-active agents in the course of their preparation.

FLUORESCENT WHITENING AGENTS. See *Optical Brighteners.*

FLUORESCENT DYES. The colorants which not only possess the color they reflect but also emit fluorescent light of the same color.

INTERFERENCE COLORS. See *Iridescent Colors.*

IRIDESCENT COLORS. A special class of nacreous pigments in which color is produced by light interference rather than light absorption.

ISCC. Inter-Society Color Council.

LAKE. Dyes precipitated on an inert pigment, such as barium sulfate or aluminum hydrate.

LAMBERT'S LAW. Absorption of light at each wavelength is proportional to the thickness of the absorbing material.

LIGHTFASTNESS. Resistance of a colorant to the degradation effect of light on the color of the object.

METALLIC FLAKE PIGMENTS. Shiny leaflets of aluminum and copper or their alloys used as pigments.

METAMERISM. The change of color of an object when viewed in different lights.

NBS. National Bureau of Standards.

NACREOUS PIGMENTS. Pigments which impart pearl luster.

NANOMETER. 1/1,000,000 millimeter.

OPAQUE. An object which scatters the incident light to such an extent that virtually no light passes through the material.

OPACIFIERS. White pigments, used to make plastic white or opaque.

OPTICAL BRIGHTENERS. A special class of essentially colorless fluorescent organic substances which absorb UV radiation of about 3000Å and emit this energy as visible radiation.

PLASTICIZERS. Low-molecular weight organic substances used to make plastics, especially PVC, more pliable when mixed therein during fabrication.

PIGMENTS. Synthetic or natural inorganic or organic colorants, insoluble in water or other solvents.

PLATE-OUT. Strongly adhering deposition of pigments or other additives on machinery parts during the processing of plastics.

REDUCED PIGMENTS. Mixture of pigments and an inert substance.

RHEOLOGY. The study of the flow of materials.

RESIN. A term loosely used to refer to plastic materials in general; more specifically, the polymers derived from natural sources and those polymers which can be cross-linked.

SPECIAL EFFECT PIGMENTS. Pigments used primarily not to impart color but other appearance characteristics, such as opacity, brightness, fluorescence, metallic reflectance, etc.

SPECIFIC GRAVITY. A measure of heaviness. Specific gravity of water at room temperature is 1.

SPECULAR REFLECTION. The incident ray and the reflected ray form equal angles to the reflecting surface. Such a reflection produces stark highlights in the general appearnace of an object.

SPECTRAL MATCH. Two colors which look identical under all conditions of viewing; also called *invarient match.*

STABILIZERS. Low-molecular organic substances used to prevent or delay the action of light, heat, and other degradative agencies on plastics when mixed therein during fabrication.

STRUCTURE OF PIGMENT. A measure of the degree of irregularity and shape of primary aggregates of pigments, especially carbon blacks.

TACK. Internal cohesive strength of a melt. The term is widely used in rubber and PVC fabricating industries.

THERMOPLASTICS. Plastic materials which can be repeatedly melted at elevated temperatures.

THERMOSET PLASTICS. Plastic materials which can be melted only once, if at all. They cross-link during fabrication, after which they defy melting.

THERMOCHROMIC EFFECT. Reversible change of color due to heat.

TINCTORIAL STRENGTH. Degree to which the colorant imparts color to a standard white pigment.

TONER. Any intense colored material used in small quantities to modify or adjust a color formulation.

TRANSLUCENT. An object which allows most of the light to pass through but which also scatters a part of the incident light.

TRISTIMULUS VALUES. The axes of a three-dimensional euclidean space in which every perceivable color can be located.

TRANSPARENT. An object which allows almost all of the incident light to pass through without being scattered or absorbed.

UV ABSORBERS. Colorless organic compounds which absorb strongly in near ultra violet region. They are used as light stabilizers for plastics.

VARIENT MATCH. Also called conditional match. The varient match may be identical under one set of illumination but differ under another set of illumination.

WEATHER RESISTANCE. The resistance of colorants and plastics to the damaging effects of light, oxygen, ozon, and other degradation agencies of environment.

Index